WHAT'S LOVE GOT TO DO WITH IT?

A N C H O R   B O O K S

D O U B L E D A Y

N E W   Y O R K

L O N D O N

T O R O N T O

S Y D N E Y

A U C K L A N D

WHAT'S LOVE GOT TO DO WITH IT?

THE EVOLUTION OF HUMAN MATING

MEREDITH F. SMALL

FOR TIM,
BECAUSE LOVE HAS SOMETHING
TO DO WITH IT.

Modern men and women are obsessed
with the sexual; it is the only realm of
primordial adventure still left to most
of us.

EDWARD ABBEY

Contents

Introduction

Last Thanksgiving, I was invited to dinner at a friend's house. I sat next to another guest, a physician. Having heard I was an anthropologist, he turned toward me and politely asked about my research. I was undecided on how to answer—should I keep my mouth shut about my years of studying monkey mating behavior and my current research project on human ovulation, or should I throw my research topics out on the table along with the mashed potatoes and the stuffing? I decided my dinner companion was an adult, and as a doctor, he certainly should know enough about human anatomy not to be put off. So I launched into an explanation. "At the moment," I said, "I'm working on ovulation— whether or not women, or their male partners, think they know when they ovulate." He put down his fork. I knew he was either going to excuse himself and turn to the guest on his other side, or plunge deeper into discussion with me. It was the latter. "And what will this information tell us? Why is it anthropology?" he asked. In answer, I explained how some anthropologists maintain that unlike the ovulation of many other primates, such as chimpanzees which have large pink swellings that announce their fertility, ovulation in human females is more concealed. This fact might have had a significant effect on the evolution of the human mating system over the past millions of years. A man who was concerned about paternity would have to guard his partner all the time to make sure she wasn't mating and conceiving with someone else. This might be one reason for an extended human pair bond. Or a woman could manipulate a man into having sex when she wasn't fertile, thereby depleting sperm supplies that he might use to impregnate other women. We chatted on, discussing the possible roles of men and women in shaping how we mate today. Meanwhile, the word "ovulation" seemed to float down the length of the dinner table, tap each guest on the shoulder, and land right on the half-carved turkey. I suddenly realized the entire table had grown silent. All the guests were turned in my direction, listening to this monologue on the mating behavior of our human ancestors. As I saw it, the only thing I could do at that point was open the topic up for discussion.

"Do you know when you ovulate?" I asked my hostess. She replied, "Why, yes, of course." Each woman in turn said "yes" or "no," or that she hadn't thought much about it before. One young woman wasn't sure exactly what ovulation *was*. When we explained, she thought that perhaps she felt something halfway through her cycle that might indicate her ovulation. I didn't have the courage to ask the male guests if *they* knew when their partners ovulated, but the question was on my mind. As an anthropologist, I'm primarily interested in the evolutionary history of this question, but I'm also curious how this biological event might subtly affect current relationships. I wanted very much to ask the couples at the table if knowing when the female partner ovulated had ever altered their patterns of sexual intercourse in any way over the years. Did they use knowledge of ovulation to avoid pregnancy, or perhaps achieve it? Could I bring this up in polite conversation over pumpkin pie? At that point, the decision was taken out of my hands when the subject was gracefully changed by my host, who took a quick glance at his oldest son and commented, "If women really knew when they ovulated, *some* people might not be sitting around this table."

The events of the evening illustrate two points to me. First, most people don't always know how their sexual biology works. Our knowledge of how babies are made is mostly connected to methods of contraception—and many of us don't even know how that works either. And sex, the process by which we make those babies, is just as mysterious. I'm sure most of the guests that evening wouldn't have been able to explain why they wanted to have sex at all, or why they were sexually attracted to one person rather than another. Second, that evening once again showed me that everyone is fascinated by anything having to do with sex and reproduction. With the simple utterance of the word "ovulation," for example, I was able to derail an audience of twelve hungry people who should have been concentrating on the dark-meat light-meat decision into thinking about their own reproductive biology. And everybody had something to say. I can't help but wonder what the reaction would have been had I brought up the topic of orgasm.

This book is about human mating, or sex, because we're all captivated by the subject—and yet totally confused about its origins. By definition, the book also touches on the human reproductive process, because making babies is the ultimate *biological* consequence of sex for humans, as it is in any sexually reproducing species. But since our particular species is not just motivated by inner biological drives, I have to look at sex as part of our culture as well.

It's odd that we know much about the world around us, and so little about our sexuality and our reproductive biology. How can we as a species guide our lives, choose lifelong partners, make important decisions about having children, or be responsible adults if we don't comprehend our basic sexual biology? To know ourselves sexually, it would seem, is to understand one of the most compelling feelings in life—the drive to copulate. More confusing yet is the fact that we are culturally drowning in a topic that we barely comprehend. In Western society, sex is thrown into our sensory field at every turn—photographs of young men in underwear are spread across several magazine pages; long-legged models lounge on shiny new automobiles to entice us to spend our hard-earned wages; sexually explicit magazines fill the newsstands to catch our passing eye; stand-up comedians know they can always get a laugh with an off-color line; and television scripts spike their dialogue with sexual innuendo to keep us from heading for the fridge. Our modern media, some suggest, is synonymous with sex. In a sense, the media is only taking advantage of the conflicted place of sexuality in our culture and our daily lives.

I believe that one of the major reasons we latch on to racy stories and sexual images is that we use them, consciously or unconsciously, to measure ourselves. The photographs, the stories, and the headlines are comparative devices for an important question—am I normal? In this the media has been more useful than it probably intended. Our sexual thoughts and acts are so private, so rarely discussed in public, that when popular culture displays sexuality in our magazines and across our television screens, we can't help but compare our lives to those images. Because our particular culture with its prudish Victorian heritage allows for little discussion of our sexuality, even among close friends, we are mesmerized by public displays. The end result is confusion. Somewhere in the unconscious, the same place where those sexual thoughts crop up, most of us know that sexuality is just part of life, as much a part of living as eating and breathing. But at the same time, our culture tells us that anything sexual has a hint of the forbidden about it. We recognize that normal, healthy people have sex, but it seems that either they're not supposed to enjoy it, or if they do, they should at least feel guilty about it.

Americans' conflict over sex makes it a difficult subject to address seriously. When I tell people I am writing a book about the evolution of human sexuality, I get all sorts of comments. Some people are extremely embarrassed and quickly change the subject. Others want to know everything I've learned so far, but they ask questions in a leering, voyeuristic sort of way. One friend even sat down and told me long stories of her sexual adventures, hoping this might get her into the book. We have more trouble discussing or acknowledging sex in a reasonable way than any other human activity, because we aren't really sure how sex fits into our lives. Sex is a personal mystery, and in our culture, forces have conspired to keep it as such, and in fact even intensify the sense of mystery. But it is possible to be comfortable with our sexuality, to attain a healthy attitude about one of our most fundamental behaviors. I believe we need to place sexuality in its appropriate context, not just the context that popular culture places it in.

My intention in this book is to look at sex from a "scientific" perspective. I follow in the footsteps of many other scientists and popularizers who have also taken a look at sex. Alfred Kinsey and his colleagues were the first to conduct a major survey of American sexual practices and attitudes in the 1940s.[1] Although many criticisms might now be lodged against their work, these surveys remain some of the most reliable tallies of human sexuality. In the 1960s, William Masters and Virginia Johnson began detailed studies on the physiology of human sexual response.[2] Others have refined their techniques or expanded their ideas, but Masters and Johnson's descriptions still provide the best overall view of how the body responds to sexual stimuli. More recently, Shere Hite allowed women and men to write about their sexuality without the constraints of scientific observers, surveys, or tallies. Her books, *The Hite Report* and *The Hite Report on Men*,[3] represent unfettered voices describing the place of sex in everyday lives. A number of surveys, most notably *The Janus Report*[4] and *Sex in America*,[5] have documented current American sexual practices and attitudes, tracking the so-called sexual revolution and the effect of AIDS on sex in America.

But still, there is much to be learned, and our society seems to be running away from this important knowledge, rather than looking for answers. We know so little about sexuality because we, as a society, have a very restricted approach to the subject. We often stop short of seeking answers to the most basic questions about sex, such as why we're so interested in the sex lives of other people, or why we jump to make judgments about what is normal or abnormal sexual practice. It's as if with sex we should be satisfied with superficial answers. I think knowing the path of physiological response, such as Masters and Johnson have described, or adding up the tallies like Kinsey and Janus, aren't enough. And so *What's Love Got to Do with It?* is an attempt to bring the study of sex out of the closet and off the academic pedestal and integrate it among other behavioral patterns that make up the totality of the human experience. After all, the sexual act, and the drive for sexual satisfaction, is both a universal feature of our species and part of our animal selves that, along with the rest of our nature, is the very essence that allowed our ancestors to survive over millions of years.

But what exactly is human nature? More important to the subject of this book, how has our sexual nature evolved? There are unique features of human anatomy and psyche that mold each individual of our species. First, we know that we are primates, a taxonomic order that includes lemurs, monkeys, apes, and humans. We primates are descendants of small tree-dwelling mammals, and most of our distinguishing anatomical features echo our arboreal heritage. For example, we primates all have excellent vision for swinging through the trees and highly flexible hands for grasping. Looking at the other primates, in fact, can be instructive, especially when it comes to broad patterns of behavior. As near relatives, what they do and how they act often echo what we do and how we act, and vice versa. At the most basic behavioral level, all primates, including humans, are highly social animals. We live in groups of various sizes and, by definition, must manipulate our social companions through competition or cooperation. We fight and then make up, we develop friendships and loyalties, form alliances and take sides. Primates also exhibit an intense concern for kin; feeding, defense, and socializing occur most often with relatives. All species of primates have highly dependent infants that require extensive care. And the wide range of mating systems seen in various human cultures can be found as well among our primate relatives. In other words, we can use knowledge of other primates to answer questions about the evolution of our own behavior. This information will determine which features and behaviors are uniquely human, and which are simply part of the basic primate blueprint that we share with other primates. I want to examine humans within our broader zoological context as animals, specifically as primates. I refer to this context often when discussing the biological basis of human sexual behavior.

At the same time, it is also important to look at humans as distinct from other species. We are thoughtful animals seeking to understand ourselves, and it's not yet clear that any other animal shares our degree of self-consciousness. We have developed an elaborate system of social interaction called culture that mediates our biological nature. This is both an advantage and a curse. We can decide whether or not to follow our biological urges rather than being overpowered by inner voices. At the same time, we sometimes forget about the flexibility of that cultural mantle. We equate our own personal behavior, or that of our immediate culture, as the way humans *are*, or should be, ignoring the diversity of human experience and behavior. Travel to a foreign country is sometimes so exotic because we encounter interpersonal practices that don't "fit" with our notions of human behavior. In Indonesia, for example, men hold hands with men, and women with women, but you will never see a physical expression of attachment between a man and a woman in public. At first, Westerners are uncomfortable watching same-sex friends holding hands and hugging because in our culture, this kind of physicality means the couple is homosexual. But clearly, there are different ways of expressing affection, and our Western traditions are only one possibility among many. The problem is that often these personal- or cultural-bound blinders sometimes keep us from integrating what other cultures do into the larger picture of human behavior. To counter this, it is important to learn about and acknowledge the practices of other cultures. This more global perspective is important particularly when addressing the basics of human mating behavior. What is typically "normal" sexuality may vary widely when refracted through the prism of other cultures.

In *What's Love Got to Do with It?*, I examine the cultural part of human mating first and foremost from a Western perspective. This is the culture I grew up with and know best. But wherever possible, I include comparisons to other cultures as well. Desmond Morris, in his groundbreaking book about the human animal, *The Naked Ape*,[6] pointed out, with a good deal of accuracy, that comparisons with isolated smaller cultures may not be of much use in understanding the larger picture of human nature. Such "dying" cultures are the losers in the human genetic pool, Morris maintains. He feels we should look instead to successful cultures in order to see how basic human nature has played out. In one sense I agree, but I also think that isolated or marginal cultures help us understand the flexibility of human sexuality. And so I have included how other peoples express their sexual urges when I feel it adds to the discussion. This flexibility in sexual practice is surely one of the more interesting aspects of the human species. By looking at other cultures, we set our own behavior into a larger comparative framework with others of our species. We discover that there are universal features of our sexuality

—concern with sex, orgasm, the search for several partners—and that there are differences. For example, most Western cultures consider male homosexual acts abnormal, or even abhorrent. And yet in many tribal groups in other countries, young men must engage in some form of what we would call homosexuality before they can fully enter the life cycle stage of manhood. If set into a framework of flexible male sexual behavior, neither attitude for or against male homosexuality appears morally right or wrong—the act has a cultural context that defines its acceptance. And other cultures teach us much about the place of sex in society. In some groups, women are free to have extramarital affairs, while in others, women are sequestered from interested male eyes. The balance of power between men and women often determines who controls the culture's sexuality. Generations of Western women, for example, have been influenced by a male Victorian ethic that says women do not enjoy sex as much as men. Women had to experience a sexual revolution in the 1970s to discover what female sexuality was really all about. This is only one familiar example of how societies influence how we view our sex lives, and it is imperative that we filter through those layers of culture and discover the common sexual nature that underlies those societal constraints. In that sense, we are the cultural persona seeking the biological animal inside.

To understand what makes us tick as sexual animals, I have to begin at the beginning, about 3.5 million years ago when the first human-like creatures emerged in Africa. Chapter One focuses on the evolution of human sexuality over the intervening millions of years. How can we possibly understand human sexual behavior without knowing its evolutionary history? After all, our mating habits evolved over time, and there are evolutionary reasons for their persistence. Throughout our ancestry, people of the opposite sex have fashioned some sort of tenuous bond that provides a dependent human child with a safe and nurturing environment. But the pair bond, the so-called monogamous relationship between men and women, is a shaky temporary compromise at best. In fact, our evolutionary history shows that a monogamous system is relatively recent, consistently violated, and as our present behavior shows, may be on the wane.

Chapter Two brings us into the present, and a very sexual present it is. Humans have disconnected sex from reproduction, and use sexuality for all sorts of reasons. The drive to have sex can be best explained at its most basic level as a means to pass on genetic material. But beyond that, there is a great deal of biology involved in how that drive is manifested, utilized, and fulfilled. Our bodies have a series of mechanical responses to our sexual whisperings. This chapter traces the sex drive from its psychological impulse to its pulse-raising climax.

The act of sex and its implications for each sex is further explored in Chapters Three and Four, where I look first at females and then at males. It is not my intention to provide a physiology text for men or women. What interests me are old notions about our sexuality and how those notions are being altered by science. Men are no more motivated to have sex than women are. Properly stimulated, women can reach orgasm as quickly as men. Menstruation may not be a "curse," and sperm have a life of their own. What we know about human sexuality today is different than what was known even twenty years ago. Women in Western culture are more sexually liberated today and we have more knowledge about how the female reproductive process works. Men, too, have undergone a liberation of sorts. We now know that not all men are driven to have sex at every turn; there is flexibility in their impulses and variation in their responses.

The sexes are rejoined in Chapter Five because our biological heritage pushes many, but certainly not all, of us to find a suitable mate and put up with him or her long enough to conceive an offspring. And we must do this within a complex framework of a particular family system, social group, economic and political system, and culture. In Chapter Five I turn to the recent explosion of research on human mate choice, mostly by psychologists and sociobiologists. This work dramatically demonstrates that we don't pick our mates at random—and we certainly don't always make the best choices.

No current volume on human sexuality would be complete without a discussion of homosexuality. Is it a lifestyle choice? A biological imperative? A psychological aberration? In Chapter Six I examine how science has added to the politicization of homosexuality. Since homosexuality is a universal feature of our species, it is time to understand it as part of human nature.

Disconnected from the hormones and genes that guide most other animals toward reproduction, our sexual behavior has flowered into a vast array of possibilities. More often than not, our copulatory acts have nothing to do with reproduction. And we have excelled in expanding the category of "sex" into something more wide-ranging, more inclusive than most other animals. We descended from a savanna-living bipedal ape and find ourselves working in factories or sitting in front of computer screens. We no longer scavenge for food but shop conveniently in super-markets or eat what fast-food chains offer up. We still sometimes have sex as a means of reproducing, but our sexual lives, too, face a future of transformation. Babies can be conceived *in vitro,* sperm from anony-mous donors is available, and it's now possible for a woman to conceive even if she's menopausal. There's no doubt, too, that our current sexu-ally transmitted plague, HIV infection, is having a dramatic effect on how individuals conduct their sexual lives. Standing over each bed is now the face of possible disease and death. This is a new influence, biological in origin and cultural in transmission that must be accounted for when searching for the sexual nature of human beings. There are other less serious cultural changes that could affect the future development of our mating patterns—dating services, telephone and computer sex. Taken together, these influences suggest that the mating habits of our descen-dants may be quite different from the sex of our ancestors. The final chapter of the book then looks ahead to human sexuality in the future, as best we can see it from here.

WHAT'S LOVE GOT TO DO WITH IT?

CHAPTER ONE

The Sexual Animal

Everyday Sex

My favorite red light district in the world lies along the Oudeizijdsachter canal in Amsterdam. I first went there at the age of twenty-three, as a backpacking youth on the usual European tour. My traveling companion and I were walking to our hostel when we paused in front of a shop selling sex goods. We stood there for a few minutes, pointing to various objects, trying to figure out exactly what someone might do with those things. We had a great time. Since then, I make it a point to wander into any foreign city's red light district during the day; a view of the so-called underbelly of a city adds a certain color to my memory of the visit and adds contrast to the finer points such as museums and churches.

Amsterdam is my favorite because walking down the street next to the canal, day or night, I always feel as if I've come to take a pleasant stroll through some charming old world neighborhood rather than a center for sexual products and services. There are, of course, the usual bars with neon signs announcing a menu of sexual pleasures in the international one-word lingua franca of sex—"Live," "Hard," "Topless," "Girls! Girls! Girls!" and the standard marquees with photographs of naked women with freakishly large breasts and seductive pouts. But as in the rest of the Netherlands, the streets are immaculately clean. Windows are sparkling, as if they'd been washed twice that day, and hung with lacy curtains reminiscent of a Vermeer painting. And oddly enough, there are no women lounging in doorways here, no hand gestures or sultry invitations to a few moments of sex down a convenient alley. Instead, the prostitutes sit in first-floor picture windows talking to each other, knitting, or just looking out the window at the passing crowds. Of course, they have on the traditional uniform of whores—minuscule shorts, boots, too small brassieres, and big hair—and they look as bored as the prostitutes of Paris, New York, or Bangkok where I've also wandered. But somehow this place looks more casual, more homey, as if these women were simply hanging out on their own front porches, waiting for something interesting to happen in the neighborhood. The street crowd is different here too. It's mostly male, but there's also a large contingent of women of all ages looking at the shop windows or passing through. Sometimes families saunter down the street, often pausing to check out a shop window. They point at the goods for sale—giant dildos, explicit videos, and objects that can only be identified by those who have a taste for whatever that contraption might be. Kids point and parents laugh as if they were looking at a joke shop display instead of the odd accouterments of sex. Surely there's an undercurrent of drugs and exploitation in this district, but the surface is pleasant and nothing to be ashamed of; it's even friendly. What I like is the feeling that it's not only acceptable to be a prostitute here, it's also acceptable to be a naive tourist out for a view of sex or sexual accessories for sale. Other red light districts are more sinister. In Bangkok, for example, I would never venture into the crowded streets of Patpong alone. There's something menacing about the pounding lights and loud music. The lovely Asian women coated with eye makeup seem trapped and pitiful as they invite passersby into clubs. Times Square in New York is also rather frightening, not only because its sexuality is hard-edged, but also because New York's violent atmosphere seems closer to the surface. Who knows whether or not the man or woman in leopard skin tights is going to try to sell you sexual satisfaction or rob you. But even these areas share something with the more congenial Amsterdam district, and it's the communality that draws me to any city's red light district for an objective look around. Past the neon lights, beyond the music, right next to the woman in three-inch stiletto heels, is normal urban life. These supposed pleasure spots with their exaggerated signs of sex are also working blocks of a city. Sex isn't the only thing for sale, or the single activity of any red light district. Cafes and bars line most of the streets so that tired sex shoppers and others can get a bite to eat or pause and have a drink. Nor is everyone there to buy sex. Some people live in the district; others are just passing through on their way to or from work and they see no reason to avoid this quarter just because of its business. Times Square in New York, for example, is a sex district, theater district, commercial center, and tourism mecca. One can just as easily buy a daily newspaper, a sandwich, or do laundry in a red light district as anywhere else in a city. Look more closely and you'll catch a glimpse of people in the apartments above the sex shops, people who get up each morning, go to work in more traditional occupations, come home, watch TV, and raise their families. Sex may be the most visible item on sale here, but all around, the usual urban activities go on as well.

In that sense, the place of sex in red light districts in cities mimics the place of sex in our culture. In both cases, sex is one part of a larger complex of life, and in both cases, sex is also set off in a world of its own. It's emphasized and exaggerated by advertisers, and repressed by certain political or religious factions. In the same way, we do a similar sort of sexual segregation as individuals when we treat sexuality as distinct from the rest of our behavior. No other normal human behavior is subject to such scrutiny and pressure. In fact, I think most cities have integrated red light districts into their tourist maps much better than most people have integrated sexuality into their persona.

The complication with sex is that we often act as if humans are different from other animals, as if our sexuality evolved for our singular plea-sure rather than as a biological means to make more of our own kind. We have separated sex from reproduction and hung it with trappings of culture. Because we are able to elaborate sex beyond its procreative func-tion, we seem to have lost sight of who we are as sexual animals. As a result, sex has become something not part of our selves, or our biology, but some disconnected feature not wholly human. We look from a dis-tance at pictures of sex in magazines and acts of sex in films and think of them not as patterns of behavior that evolved as part of our nature, but as something to be hidden, whispered about, perhaps be ashamed of. Sexuality surrounds us, but it is not necessarily part of us in an easy way. And yet underneath all the rituals of marriage, beyond the pop songs of love and romance, somewhere behind the porno shops and trashy maga-zines, lies the sexual animal within. This creature evolved over at least 4 million years and only recently, as little as 100,000 years ago, gained what we now call "culture." I suggest we turn our backs to the red light districts, toss out the advertisements with the suggestive naked women, peel off those layers of culture, and look to our past for clues to our sexuality. Only in this historical and evolutionary context can we begin to understand and properly position the place of sex in our lives.

Our Ancestors, Ourselves

We are a lucky species. We're not only intelligent enough to ponder about our past, but we've also evolved certain tools that aid us in a search for our ancestry. The same brain that makes us ask the question "Where did I come from?" has the ability to sort out the pieces of our puzzling past and then assemble them into a reasonable scenario of hu-man evolution. This story is not set in stone; it, too, is an ever evolving history that changes as scientists discover new puzzle pieces. For exam-ple, only twenty years ago anthropologists believed that humans evolved from a single ape-like ancestor, going through various stages in a straight-line fashion. Today, after the discovery of many more pre-human fossils, we know that the human lineage was but one of many human-like forms that roamed Africa over a million years ago. Our history is not one of direct descent but one of branches of a tree on which we had many near, but now extinct, cousins.

IN THE BEGINNING

The Miocene epoch, starting about 25 million years ago, was a time of global cooling. Although forests covered most of the earth at that time, they were slowly retreating and being replaced by savanna grasslands.[1] If we concentrate just on primates, our own order, there were several types of prosimians in Africa and Asia, and many species of monkeys, creatures that eventually evolved into the seventy or so species of monkeys alive today. The Miocene was also an age in which apes flourished. Today there are only four species of apes remaining on earth—chimpanzees, gorillas, gibbons, and orangutans—and all are highly vulnerable to extinction. But during the Miocene, at least twenty species of apes lived in Africa, Asia, and Europe.[2] Within one of those ape lineages we find our beginnings.

Between 14 and 8 million years ago, during the forest-shrinking Miocene epoch, humans shared a common ancestor with what we know today as the chimpanzee.[3] This is not to say that we are descended from chimpanzees—we're not. But long ago, one species of Miocene ape separated into two separate lines. One line eventually split again into common chimpanzees and another kind of chimpanzee, bonobos, and the other line eventually evolved into *Homo sapiens,* our own species. Paleontologists aren't sure exactly what happened at the human-ape split. Only bits and pieces of fossils dating from 14 to 4 million years ago have been recovered, and these bits tell us almost nothing about why one line of ape evolved in a different direction from the others.[4] Scientists know only that 3.5 million years ago, a very distinct creature lived in Africa that had hallmarks of humanness. The species, typified by the fossil remains known as Lucy found in Ethiopia in 1976, is called *Australopithecus afarensis.* These creatures were rather small: the females were about three and a half feet tall, males perhaps five feet tall. Their teeth were rather human-like, although some still retained the larger canines of apes. Curiously, they also had curved hand and feet bones, which suggests either that they spent some time in the trees, or that it hadn't been long since tree climbing was important to their way of life. Lucy and her contemporaries had a cranial capacity much like their ape ancestors, about the size of a modern chimpanzee at 400 cubic centimeters. Given their body size, this is still a big brain when compared to other mammals, but nothing like the size of a modern human brain, which is about 1200 cubic centimeters. What defines *afarensis* as human is their locomotion. For the first time in the fossil record, a mammal used a bipedal gait to get around. We know this from the shape of Lucy's pelvis; it's short and broad, designed to support large buttock muscles like a modern human pelvis, rather than long and thin like a chimpanzee pelvis. In addition, their knee joints had evolved to bear the weight of an upright walker. But *afarensis* left an even more obvious clue to their pedestrian way of life—a series of footprints laid down over three million years ago in Tanzania that clearly show the swinging gait of a habitual biped.

What made these creatures stand on two legs is unknown. At first, anthropologists speculated that standing on two legs was important for freeing the arms to carry things, such as tools, children, or food.[5] Today, many anthropologists approach the question of our shift to bipedalism as they would any major change in a species' history—they look for the reason natural selection might pressure individuals to change their ways. From that perspective, there must have been some sort of environmental strain that forced *afarensis* or its immediate ancestors to adapt a new way of getting around.

There is good evidence that our ancient ancestors had a major environmental crisis on their hands—the forests were shrinking; since they were most likely forest-living animals, their world was literally disappearing before their eyes. It may be that one species of ape experienced the change in environment as an opportunity; members of that species moved into the open grassland and were able to exploit new sources of food not used by those who stayed behind in the trees. But the move onto the savanna came with a price. A forest provides a reasonably even spread of food if you rely on green matter and fruits. Once on the savanna, where such food appears more seasonally and is located farther apart in groves or clearings, our ancestors would have to travel long distances to find enough food throughout the year to accommodate their large body size.[6] Under these circumstances, bipedalism would have proved a more efficient means of locomotion. For a large-bodied ape, walking on two legs for long distances at a steady but unhurried pace uses fewer calories than covering the same distance on four legs.[7] In this way, the direct ancestors of *afarensis* might have made the transition from the forest to the grasslands that sent them wandering on two legs from one grove of trees to another, passing through miles of open savanna as they scavenged for food. By 2 million years ago this transition was virtually complete and a new lineage that eventually led to modern humans was well established.

THE DIVERGENT HUMAN TREE

Whenever environments change, species either adapt or die. Those that do adapt tend to experience rapid changes in the way they look and behave—they're essentially running one step ahead of extinction—and there is often a flowering of new species during these periods.[8] And so it was in our history. Today we see only one kind of human all over the earth, modern *Homo sapiens sapiens*. But between 2.5 and about 1.5 million years ago, there were at least five different kinds of human-like creatures in Africa.[9] This list now includes four Australopithecine species who were fully bipedal but had very small brains, like *afarensis*, and a larger-brained group called *Homo habilis*. It's probably from *habilis*, which first shows up in the fossil record about 1.8 million years ago, that the modern human lineage evolved.

From this point, our ancestry looks reasonably familiar. The large brain rapidly becomes even larger. Only 750 cubic centimeters in *Homo habilis*, brain size almost doubles over the next million years as our line is traced through *Homo erectus, Homo sapiens neanderthalensis,* and various transitional species to modern *Homo sapiens sapiens*. Why we have big brains is unknown, although there are all sorts of suggestions. Obviously the use of tools requires brain power, but tools appeared in our history long before we had a large brain. Language might explain the need for a big brain, but human language is a relatively recent development and is probably a product of our complex brains, rather than its catalyst. One possibility is that as social beings we needed larger brains to run our interpersonal interactions.[10] Social intelligence, that is, keeping track of relatives, enemies, and friends, could be the force that selected for a rapid increase in the size and complexity of the human brain.

In any case, we put those additional neurons to use early on in our history. There's evidence of high intelligence and puzzle-solving skills in the stone tools found with *Homo habilis* about 1.8 million years ago. These crude blocks of rock could have been used to smash bone or skin animals. Tools probably helped the species exploit a wider range of animals by making hunting and scavenging more efficient. Tool manufacture and development probably also aided in processing vegetable matter. It's possible that without tools and the things they make, later humans might not have been able to move out of Africa into more difficult climates around the globe.[11] And so the history of our species is distinguished by bipedalism, a large brain, and the ability to make and use tools and other cultural artifacts that help us alter the world to fit our needs. Left to our imagination are the other facets of our lives, such as how we interacted with family and friends, and how we ran our sex lives.

What We Make Is Not Who We Are

Although fossils and stone tools outline the story of our biological and cultural history, stones and bones don't really tell who we are. Unfortunately, behavior doesn't fossilize. Sexual behavior, for example, is almost impossible to find imprinted in any piece of old bone. And so we

are left wondering who were those members of our heritage, walking across the savanna, eating fruits and bits of meat, and having sex?

THE NATURAL BIOLOGY OF HUMANS

If we think of humans as only one species of animal among many, they merit the following description: They are a large mammal with a large brain; bipedal with long legs; and the males overall are larger than the females. Males have hair on their head, face, under the armpits, and covering the testicles, but are otherwise relatively hairless. They have large penises and small testicles, which are easily observed from a distance. Females have fatted breasts, hair on their heads and armpits and in the pubic region. As a primatologist used to watching monkeys, I might notice that female humans don't have any outward displays of ovulation like the red butts of chimpanzees and baboons. But human females do display the end of a cycle with a visible bleeding from the vagina, and these bleedings stop in middle age, which means females have a restricted reproductive life. When females do conceive it's usually one infant at a time. Human babies are extremely dependent from the moment of birth and nurse for several years, and both parents seem to be involved in caretaking. Unlike most mammals, these offspring remain dependent on their parents for food and shelter through the juvenile stage, and stay close even as adults.

This is the kind of description I would give of a new species of monkey just discovered in the forest. I note these particular features because, familiar as I am with other primates, these are the details that distinguish us from apes, at least in the physical sense. Some of these features are surely involved in human sexuality, and as such, they are our only physical clues to the natural sexual biology of our species. The way people behave sexually in everyday life now doesn't necessarily imply our actions are only the result of what nature compels us to do—culture and society often push and mold us in all sorts of directions unknown to our ancestors. But in some ways our body parts still speak the language of natural selection—certain features of our species are there because long ago, as

well as today, they were favored by natural selection to help us pass on genes. And so anthropologists have spent great effort trying to tie these features into a coherent story of human evolution. Yet so far no clear picture of our sexual makeup has emerged.

A SENSE OF COMMUNITY

Even if we aren't sure how our ancestors behaved sexually day to day, we do know they were social animals living in an interactive community. In fact, almost all primates are social animals. Unlike most other mammals, primates typically live in groups and pay close attention to each other. They touch, groom, sit close to each other, and interact constantly. We reason that our ancestors did as well, and there's also some paleontological evidence to support this idea. In 1978, the group that discovered Lucy and named *Australopithecus afarensis* found a cache of bones that represent at least thirteen *afarensis* individuals. This community of fossils might not have lived together, or even died together, but it would be quite a coincidence to unearth several individuals in the same spot long after they died if they hadn't somehow been in association. The question of community is important to our understanding of our sexuality because, like all animals, our mating system is embedded within our larger social system. Group living and the members who comprise that group define the mating opportunities. A large group means there are many possible mating partners, while a small group limits the choices to a few. Given the reasonable guess that our history is one of group living, what kind of group might our species have lived in?

There are several possibilities, all of which are based on a species response to the question of food. At the most basic level, every individual is biologically programmed to pass on genes. To do that, an individual must stay alive, have sex, and bring up infants to a reasonable age. For males, this means avoiding predators and using every opportunity available to have sex and place sperm into fertile females. For females, especially mammals, having sex and conceiving isn't as much of a problem; males tend to congregate around females, the more limiting reproductive resource. Although females must also avoid the jaws of predators,

their main concern is getting enough food in the long run to nourish a pregnancy and nurse an infant. Females, according to this theory, distribute themselves according to the distribution of food. How they do this is constrained by the ability to first locate, gather, and digest that food under the watchful eyes of predators. Males, for whom food is not quite as important, in the reproductive sense, as finding females, will group according to how females are grouped.[12] In other words, if we want to understand why baboons live in large communities of up to one hundred individuals or so and gibbons live in small family groups, and to extrapolate that reasoning to the natural history of human groups, we have to look first at females.

Baboon females are small animals that forage on vegetable matter evenly spread over the savanna. Females don't need to defend a specific area, but need protection from predators as they forage across the savanna. As a result, they live in groups with their sisters. Male baboons would prefer to sequester a few females alone, but since they can't manage that with such a large group, other males tag along too. The result is a large community of both males and females. Female gibbons, on the other hand, aggressively defend a specific patch of forest that is rich in food from intruders, including the female gibbon's sisters, and only one male is allowed to live with her in mutual defense of the area. But most of the male's time is spent not so much defending their food as it is defending the female herself from other interested males.[13] The result is a pair-bonded couple with a few offspring, a family group of sorts. In each instance above, the social system, and by implication the mating system, is related to how animals get their food and their mates.

Perhaps our ancestors, too, lived in large groups like baboons, roaming the savanna in packs of many males and many females, with easy sexual access to each other. The only problem with this scenario is that humans are relatively large animals. As a group dependent on foraging and hunting, we would never have gotten enough to eat; groups of early hominids surely couldn't move fast enough or travel far enough to ensure that everyone in a large group got fed. Our near cousins, the chimpanzees, solved this problem by splitting into what primatologists call a fission-fusion society. They come together on occasion, but forage mostly alone.[14] More likely, our ancestors lived in smaller, somewhat

permanent, groups ranging from one male living and mating with a small group of females, to a family group with a monogamous pair in charge. And herein lies one of the major controversies of our species—are we a "naturally" monogamous species with a long history of faithfulness, or a species with a more checkered past?

The Natural History of Human Mating

Marriage is a human universal. In all cultures, men and women pair off, two by two, establishing a family unit. They always do this with public ceremony, ritual, and tradition. On the surface, marriage provides easy sexual access for both partners, but the union is not strictly a private sexual matter. Public marriage ceremonies, more than anything else, demonstrate in all cultures that marriage is intended to formalize the system in which most children are born and raised; marriages also represent the joining of extended families that can translate into alliances. In addition, marriage can be a political expedient. In a sense, sex is probably the last thing a marriage guarantees. But underneath all those culturally imposed trappings, people still choose to enter into legally sanctioned pair-bonds even without family or political pressure. Because marriage is a feature found in all cultures, and because many people feel compelled to marry and establish a family life regardless of expediency, it's reasonable to suggest that it might be the "natural" biological state to pair up with someone for life. From this view, we seem biologically like a monogamous species. Or are we?

THE EVOLUTION OF MONOGAMY

Monogamy, as a mating system, not a sexual practice, is rare in the animal world. It is formed through a tenuous compromise between two

wary partners that comes about only under very special circumstances. There has to be a compelling reason for a male to stay with one female and an equally compelling reason for the female to let the male remain with her at all. Monogamy can evolve, for example, when a female lives alone but needs a male to help her defend a food source.[15] It can also evolve because of her need for help with infant care. But a male will invest in these babies only when he's sure the offspring are his; why should a male waste his paternal time unless he is assured of paternity?

If we assume for a moment that humans are "naturally" monogamous, was territory or paternity at stake when our ancestors moved toward a monogamous way of life? We know that our pre-human ancestors roamed the savanna in search of widely spaced patches of food; chances are, they weren't particularly territorial. It also seems reasonable to suggest that ancient human females weren't traveling the savanna alone and thus available for sequestering by one male. More than likely, as I mentioned, these humans lived in small groups of several adult females and probably a few males. Defending a particular territory wasn't much of an issue. A more likely thesis is that about 4 million years ago, pre-humans already had highly dependent infants that needed care from more than one parent. Infant dependency, then, might be the major pressure that formed our particular mating system.

THE HIGH-MAINTENANCE HUMAN INFANT

Biologists classify all infants according to a baby's ability to survive on its own. Those who are alert and independent at birth, like deer which run soon after birth, are called precocial. At the other end of the scale are babies which need to be carried, protected, and constantly fed. They're called altricial infants.[16] Present-day human infants have a gestation of 267 days and usually nurse for several years. In some ways, this schedule is in line with other large mammals. For example, black rhino infants gestate for 475 days and nurse for 5 to 6 years; bottle-nosed dolphins are pregnant for 360 days and nurse for about a year; gorillas are pregnant for 252 days and babies nurse for 14 months.[17] In addition, human babies have a brain that amounts to about 12 percent of their body weight, which is just about the same as in other mammal babies. The difference is that the brains of human infants grow at a pace after birth that outdistances any other animal on earth.[18] This means that although human infants are born with the appropriate-sized brain for their large mammal bodies, that brain is really unfinished when you consider what it will be like as an adult brain. Human infants are thus extremely altricial and require intense and extended caretaking because their brains can't yet command the muscle movement and thought processes found in more precocial animals.[19]

Why the human lineage developed along the lines of dependent altricial infants is unknown. Anthropologists speculate that natural selection, in this case, favored animals with large brains and high intelligence. Selection for larger brains resulted in newborns with unfinished neural networks. This in turn meant that parents would have to tend to these helpless infants. But time caring for helpless infants could also be used to teach complex social skills. This evolutionary direction resulted in highly dependent young ones that need years of attention. In the end, humans, both male and female, can only pass on genes, that is improve their reproductive success, if they cooperate and invest heavily in each infant.

THE HUMAN PAIR-BOND

Because of the need for intense and extended parenting of children, most anthropologists suggest that a pair-bond with two attentive parents is part of the human ancestral, and therefore genetically molded, mating pattern.[20] In other words, we mate two-by-two today because this type of mating pattern was selected over time as the best context for bringing up infants. The proposed, and unproven, scenario goes something like this: The human infant developed a large brain and had to be born before its neural time. It was therefore highly dependent. Ancestral females couldn't manage all the child care on their own, and so they needed a partner. The only way they could keep a male close at hand to help with raising the children was to be available to have sex with him all the time. And so females shed any outward signs of their fertility so that males would be fooled into thinking they were always fertile. But this also presented a problem. If males knew when females ovulated, they could guard each one from other males during the important, that is fertile, times and look around for other females the rest of the time. But since there were no outward signs of fertility, and females were available for sex regardless of their fertility, males had to form exclusive and guarded pair-bonds with particular females to keep them from other males. As a consequence, males were more assured of the paternity of a partner's infants and might help out in infant care. The female, by being always sexually available, gained paternal care for her infant, and perhaps an increased food supply, such as meat. In this scenario, the nuclear family is born in the roots of sexuality, and females are the initiators of this system because they produce those dependent infants.[21] The only problem with this scenario is that it's based on feet of clay.

FOCUS ON FEMALES

Anthropologists have studied, examined, speculated, and hypothesized why women have the particular sexual biology that they do by contrasting humans with chimpanzees, assuming that humans have evolved away from some primal chimp-like stock. After all, it's human females that have alluring breasts, and have developed the ability to have sex all the time, unlike their chimpanzee cousins, which sport huge genital swellings during heat that broadcast fertility. It's true that human females don't have estrus, or heat periods, like other animals and are what the researchers call "continuously receptive" to sex.[22] But this doesn't necessarily mean that ancestral women lacking a clear estrus were able to snare men by offering them the possibility of sex on demand.[23] As every sexually active person knows, no long-term relationship is held together by sex. Nor does sex seem to be the glue that holds together the lives of our pair-bonded cousins in the primate world, either. For example, gibbons, the small apes of the Southeast Asian rain forest, live in a pair-bond and rarely have sex at all. And yet bonobos, the chimpanzees of the Zairian forest who have sex at every turn, are highly promiscuous and do not pair-bond. Sex, therefore, seems to have limited influence on the kind of bonds primates establish.

It's also important to question the concept of estrus, or lack of same, in this discussion of human mating systems. The idea of human females forever receptive to sex refers to a classic animal behavior term—estrus— that is often misused. For female animals, sexual behavior usually is something distinct from all other behaviors. Most of the time, female animals are not interested in sex and certainly don't seek it out. Only at certain times, pushed by hormonal changes that have evolved to facilitate conception, do females become interested in mating. They usually experience a period of heat surrounding ovulation, a time when they become especially attractive to males, will be receptive to mating postures, and often seek out males as well, a condition called proceptivity.[24] For an animal other than humans to be fully in heat or estrus, she must experience all three states of attractive, receptive, and proceptive for the mating process to be carried out. This kind of biological motivation is particularly necessary for solitary and nocturnal animals who would otherwise have trouble finding partners.[25] Since human females don't have the physiology for distinct estrous changes, none of these categories apply to them.[26] In fact, human females are almost always attractive to males. Perhaps not all women are attracted to all men all of the time, but in general, women don't appreciably change their behavior or physiology to attract the interest of men. Human females also don't have particular times when they are receptive or not; the term doesn't apply because women aren't hormonally induced to maintain a particular posture for the sex act. And proceptivity applies to women in only a vague way. In the animal behavior category of proceptive, females seek out males and make it clear they are interested. Human females certainly do this, but as yet, no one has shown that female sexual assertiveness *only* occurs with hormonal fluctuations. In other words, the defining terms of estrus, attractiveness, receptivity, and proceptivity, just don't apply well to human females.

This lack of a clear estrous period can be seen to a lesser degree in our primate sisters. Although monkeys and apes do have estrus, these periods are ill defined. A female chimpanzee or macaque will clearly show all the signs of attractiveness, receptivity, and mate seeking, but this behavior stretches far beyond the window of opportunity relative to conception. Even though some monkey females develop a big pink swelling on their backsides when they are in heat to indicate visually that they are ovulating sometime soon, swellings can go on for weeks.[27] Estrus for these females is a continuous state that turns on the appropriate mating responses for long periods and not just a quick indicator of fertility. Also, several species of other primates lack any clear signs of estrus, just like human females.[28]

So while women are always attractive, not necessarily proceptive according to their cycles, and need not be receptive to have sex, they are sexually flexible, meaning that women don't suddenly become sexual only when they are near ovulation. They can be sexual just about anytime in the ovulation cycle.[29] Clearly, this flexibility is a distinct feature of our species that must have had a significant impact on male-female relations. It also suggests that women, as well as men, might not be biologically designed for monogamy.

WHEN MONOGAMY ISN'T

Our notions about monogamy are in need of revision. Monogamy is an extremely rare system in nature. In fact, pair-bonds are found more often in our own order, the primates, than in any other. There are about two hundred species of primates, and about 14 percent of them live in pairs and mate monogamously. Monogamy is found throughout the primate order, including prosimians, monkeys, and among apes. This evidence would seem to offer biological evidence for human monogamy. But a study of supposedly monogamous titi monkeys in the Amazon basin conducted by psychologist William Mason of the University of California at Davis conducted in the 1960s gave primatologists their first hint that all was not as it appeared among seemingly monogamous primates as well.[30] Titi monkeys occupy patches of forest high in the rain forest canopy. They wake up each morning singing in unison, guarding their territory from other titis who might want to usurp their territorial rights. The typical titi couple spends much of the day side by side, tails entwined —surely a sign of an intimate and permanent attachment. Permanent, that is, until the female decides to unwind her tail, run over to the next territory, and check out a neighboring male. She sometimes copulates with the neighboring male and returns as if nothing had happened.

Those who believe humans are naturally monogamous point out that at least one of our ape cousins, the gibbons, live in close-knit pair-bonds where the male and female mate exclusively for life. But a recently published two-year study of gibbons in northern Sumatra has revealed a surprising discovery—a pair-bond is not always a pair-bond, and a family is not always a family.[31] It is true that gibbons spend their time in small groups of one male and one female with a few young, but this isn't necessarily a permanent nuclear family. The male might be a recent widow who has been joined by a female who just deserted her previous mate. And those so-called offspring may in fact be two young males from the neighborhood who left their relatives in favor of a reconstituted family. This study also showed that the siamang, the other type of small ape which is also supposed to be a staunch monogamist, isn't strictly monogamous either. At least one siamang female in the study area repeatedly left her usual partner and copulated with three different males in the area before returning home. Monogamy, then, at least for other animals, isn't what we thought it was.

ARE HUMANS NATURALLY MONOGAMOUS?

Although research on other animals suggests that monogamy might be somewhat different than the traditional definition, this doesn't necessarily discount the role of monogamy in human societies today and in the past. After all, there's an important evolutionary reason why humans should pair up—to bring up a dependent offspring. Monogamy, if it was selected for in our ancestors, must have therefore been driven as much by a need for males as well as females to invest in offspring. In other words, when nature opts for intense parenting, both genders must benefit when they stay together. The question remains, are humans compelled by their genes to stay with one partner?

The fossil evidence suggests that we were anything but monogamous. The story begins by comparing human shape and size to other animals'. In almost all species, it's easy to tell males from females. Sometimes the sexes come in two distinct colors, or one of the sexes may have brash ornaments such as horns or tail feathers that advertise their gender. More often, the sexes are of different sizes, with males usually, but not always, the larger sex. In 1859, Charles Darwin suggested that this dimorphism between the sexes evolved through mating competition. When males fight among themselves for access to females, natural selection will favor those who are larger or better accessorized for battle. Also, some brighter or more ornamented individuals might catch the eyes of the opposite sex and more frequently gain mates through their special attractiveness.[32] In both cases, differences in size and ornamentation between the sexes is driven by who gets to mate. Many primate males are much larger than their female counterparts. Baboon males are at least twice the size of females, as are orangutans and gorillas. Wherever there's a major difference in size between the sexes in primates, it appears in polygynous groups where many males must compete with each other for a chance at mating.[33] Males evolved large body size so they can battle with other males for access to females. But when primates mate monogamously, one male with one female, males don't have to compete with each other, and therefore males and females tend to have closer to the same body size.

Within this framework, humans are a mildly polygynous species that has evolved from a highly polygynous species.[34] *Australopithecus afarensis* was a highly dimorphic species[35]—so much so that paleontologists once thought the smaller and larger skeletons represented two distinct species. Females *afarensis* were only 64 percent of the size of males, which makes them less dimorphic than gorillas but more dimorphic than modern humans. This difference in body size is more in line with chimpanzees, which suggests our ancestors might have had a mating system like chimps'. And chimps certainly aren't monogamous. Males vie for females and females usually mate with several males during estrus.[36]

Through our evolutionary history, humans became less dimorphic.

Today, women are about 80 percent the size of men, which suggests that even though men and women might not be as different as our 3-million-year-old ancestors, males are still a bit larger than females on average. Our body size then underscores the possibility that we might be naturally less monogamous than we might like to believe. It's possible to speculate that the difference in male and female body size decreased over millions of years to its present 80 percent figure because we have evolved more and more toward a monogamous mating system. Data on the mating status of our present species, however, suggests that our bodies are right in line with our sexuality and mating habits.

MOONLIGHTING

No one would dispute that humans today do pair up, so our mating system could be called monogamy of a sort. But what kind of monogamy is this? In evaluations of the myriad aboriginal and industrial societies around the world, only 16 percent of human societies call themselves monogamous—the other 84 percent claim they are polygynous.[37] In reality, only about 10 percent of the men in these so-called polygynous societies actually have more than one wife.[38] Most marriages in all societies are made up of one man with one woman because only men of high status and wealth can afford more. More important, a marriage system is not necessarily a mating system, and humans repeatedly demonstrate that they are certainly not sexually monogamous. Anthropologist Helen Fisher has shown that humans most often divorce and change partners after four years of marriage, about the time infants become independent.[39] As Fisher and others have pointed out, humans do seem to marry one person at a time, as a rule, but they don't often stay with that person for life. In that sense, at least in marriage, humans seem to engage in serial monogamy—one at a time but more than one over time.

Even when humans do commit to a monogamous relationship by attending to the ritual of marriage, that often doesn't mean that they are sexually exclusive for the rest of their lives. Recent studies of adultery show that women and men all over the world like a new experience every once in a while. According to *The Janus Report on Sexual Behavior,* 33 percent of married men and 23 percent of married women admit to at least one affair.[40] Other earlier studies have found that from 25 to 50 percent of both married women and men have had some kind of sexual encounter with another partner during marriage.[41] It used to be that unfaithful husbands outnumbered unfaithful wives by a wide margin, but since the 1970s, women have been quickly catching up. The authors of the Janus study feel that extramarital affairs are now a significant feature of most American marriages.

A look at other cultures shows that "monogamous" Americans are pretty much like people all over the world. In 73 percent of cultures worldwide, both women and men admit to extramarital affairs.[42] Although the rate for men is always higher than that for women, in every culture where men take on a new partner, some women do too. More surprising, women act out their extramarital fantasies even though all societies punish women for their indiscretion more severely than they punish men. Anthropologist Sarah Blaffer Hrdy has suggested that women are discouraged, and sometimes severely punished, for infidelity simply because the men in power are afraid of the unleashed sexual drive of their partners.[43] If a woman isn't held back, these men think, she will run off willy-nilly and mate with any man she can find.

And there seems to be more biological evidence to suggest that although society imposes monogamy on us, we humans are biologically designed more for polygamy. In a remarkable study, two British biologists, Robin Baker and Mark Bellis, recruited men in monogamous relationships who were willing to contribute semen samples for a study by using a condom whenever they had sex.[44] The scientists discovered that when the men spent even short periods of time away from their mates, their sperm counts skyrocketed the next time they made love. And their sperm counts didn't go up just because they hadn't had sex for a while— many men masturbated while their mates were away. Baker and Bellis feel that Nature somehow knows that their absent mates might have been off doing something more sexually active than going to work or visiting relatives. In response to a system designed hundreds of thousands of years ago, men involuntarily release more sperm than average to overwhelm the potential competition and increase their own chances of reproductive success.

If men in most cultures desire additional wives and women join them in taking on lovers, and if our biology itself points to a system of mild polygyny, how can humans be labeled "naturally monogamous"? They aren't. The monogamous marriage system is something culture invented to make people stay together for life. There are all sorts of nonbiological reasons to make a marriage—reasons based on economics, politics, and morals all combine to make individuals conform to what's best for a particular society. But this doesn't mean that humans easily adapt to what might be best for the collective whole. Men and women initially agree to stay with each other and care for babies, but both sexes would really like an occasional fling if they could get away with it, and they often move on to greener pastures when the time is ripe.

The Accessories of Sex

I have in my slide collection about five hundred photographs of the rear ends of monkeys: pictures of female Barbary macaques walking through the forest with huge pink balloons attached to their behinds, of chimpanzee females with rosy bulbous rears presented to males, and of young female baboons with huge swellings being inspected by interested males. It's rare collection—who else but a primatologist studying the mating behavior of monkeys would spend so much time pointing the camera at an animal's backside? I amassed this collection because I'm interested in how female primates go about their mating behavior, and swellings, when they appear, are beacons of a female animal's reproductive state. They signal to males, and the scientists watching, that the female is near ovulation and probably ready to copulate. As the hormones of the female cycle change and initiate ovulation, tissues sensitive to those hormones fill with water and change color.[45]

What's of interest to me is how female monkeys act relative to changes in their swelling state. Do they seek out males instead of waiting for males to come to them? Are they interested in *particular* males when ovulation is close? Can I predict the father of a conception relative to who copulates during what phase of the swelling? Only twenty out of two hundred primate species have extravagant ovulation displays like the swellings on the ends of chimpanzees.[46] But the issue of advertising ovulation the way some primates do is important to the human story. Most anthropologists have assumed that human females, long ago, did signal ovulation to males, and that over time, we somehow lost this feature. This so-called loss of estrous signals has become implicated in the evolution of human sexuality and mating patterns. Those who follow this line of thought ask why humans eventually concealed their ovulation while many of their primate cousins continue to call attention to their fertility.

All sorts of theories have been concocted to explain concealed ovulation. The most popular, as I described earlier, is that ancestral females who didn't signal their ovulation were better able to use sex as a means to pair-bond with a male, and they needed that bond to help bring up dependent babies.[47] A more elaborate description of their scenario goes something like this: Humans are the only group-living species in which monogamy is the major mating system. Sexual swellings and advertised ovulation might incite male-male competition, which would undermine any possibility of a pair-bond. A better way might be not to advertise with a visual display, so that no one is aware of ovulation. Along with not advertising, females also "disconnected" from the hormones that swayed them into restricted periods of sexual interest; they "lost" estrus, and without estrus they could have sex at any time, not just during ovulation.

For males, lack of sexual estrus and concealed ovulation presented both a problem and an opportunity. Since ovulation was unpredictable and females were interested in mating potentially at any time, there was no way to ensure a female would conceive with a particular male. On the other hand, males, in theory, have been selected to mate at every opportunity, and now they had opportunities galore. A system evolved, some think, that accommodated the needs of both sexes. Females concealed ovulation but used their continuous sexual receptivity to establish a long-term pair-bond with a male who would help her raise infants.[48] Males conceded to this system because they could have sex anytime during the cycle; with the female mating exclusively with him, he was more willing to invest in raising infants because of his assured paternity. This scenario suggests that concealed ovulation was one of the forces that pushed our ancestors toward a monogamous way of life.

There are variants on this theme. Anthropologist Sarah Blaffer Hrdy, who has documented the destructive effects of infanticide in several primate species, believes that concealed ovulation is also an effective anti-infanticide strategy. When a female hides her ovulation and mates with several males, rather than just one, she shields her offspring against infanticide by males because any one of them might be the father.[49] Promiscuity and copulation during all times of the cycle is then a strategy for confusing paternity. Such matings might also gain for the female some parental care from all the possible fathers, so that she might receive food such as meat.[50]

There's the possibility that a female with no outward signs of ovulation has the best of both worlds; she could mate with one male on a regular basis but then cuckold him when she felt like it.[51] There's an even more radical possibility that ovulation was concealed not because it gave females something to parlay with but out of expediency. Biologist Nancy Burley has suggested that if ovulation weren't concealed in such a highly aware species as our own, few women would get pregnant and give birth. According to her, ovulation was concealed to fool females into improving their own reproductive success.[52]

In each of the above scenarios, our ancestral females, knowingly or unknowingly, used their sexuality to further their reproductive success, and they usually did it by deceiving males in one fashion or another. Although some aspects of these scenarios ring true—we do not in fact exhibit flashy genital enlargements when ovulating and we do have dependent infants—other parts of the standard theories reveal major holes. And the biggest hole is the assumption that human females once had swellings and somehow lost them.[53]

Our most primitive existing primate stock, the prosimians, don't have true sexual skin that could announce anything.[54] Instead, they usually communicate their estrous status through smells. Female prosimians in estrus mark branches with odiferous urine, and males often directly in-

spect females to note their condition. The monkeys of South and Central America, considered rather primitive when compared to Old World monkeys and apes of Asia and Africa, also show no swellings.[55] This seems to suggest that visual advertisement of ovulation is a recent development, rather than a primitive condition shared by all our primate relatives. Even among Old World monkeys and apes, extravagant visual displays of ovulation and the presence of specialized skin patches or swellings that are sensitive to hormonal fluctuations are seen among members of nine genera, and missing from the other nine.[56] Given that, who's to say which is the more generalized, more ancestral, condition? It's probable that there were three separate evolutions for this specialization, which originated as simple puffy genitalia, and then gradually came to involve other areas of the backside and the legs. Once primates developed good color vision, the pink and swollen bottom became a preferred signal for species relying on vision for interpreting the world around them.[57] In this case, the fact that humans do not exhibit ovulatory displays has nothing to do with a mating system and parental care. In fact, the question should be reversed—why do certain species of primates have swellings while humans and dozens of other primates do not?

Today, this seems the more appropriate evolutionary question. A swelling is a feature that requires calories—energy to maintain itself. For example, the body weight of a pigtail macaque monkey, when sporting a swelling, increases by 17 percent. A chimpanzee swelling includes more than a liter of water.[58] Swellings also interfere with everyday activities—I've seen Barbary macaque females with sizable wounds on their swellings that must put the female in risk of infection. The only reason for a species to evolve such an accessory is as a necessary signal, a signal that must surely be directed at males. But why would males, who are supposedly always interested in mating opportunities, need to be awakened to the possibility of a copulation?

The answer probably lies in the social system of the animals involved. Primatologists have shown that sexual swellings occur only in ground-dwelling primates such as baboons, macaques, and chimpanzees which live in groups—many males vie for a bevy of females.[59] At the same time, many females living in multi-male groups, such as vervet monkeys, show few or no signs of ovulation.[60] But swellings are not found in several species where one male is paired with one or a few females, such as marmosets, gibbons, and humans. In light of this pattern, it's reasonable to suggest that primate females evolve swellings as signals when they need to catch the attention of males at a distance, or to spark the interest of males who might be surrounded by other females.[61] In other words, when the female of a species is relatively assured of gaining a mate, she need not evolve specialized signals to gain sexual attention.

Where does this leave humans? While it's possible that our ancestors concealed ovulation and extended their ability to have sex throughout the reproductive cycle so that males would stay close, it's hard to imagine why males would be fooled into having sex with only one female when their biology tells them to pass on as many genes as possible with as many females as possible. It seems more likely to me that ancestral human males began to pair with ancestral human females a few million years ago as a response to the highly altricial infant.[62] For males, too, reproductive success was now a function of feeding and protecting a series of children that needed the care of more than one adult. I think it's more likely that human females never had swellings announcing ovulation in the first place because they were group-living primates with males close at hand.

Of course there's yet another possibility. Perhaps humans haven't really concealed ovulation at all. True, we don't have massive pink balloons on our rears, but who's to say that an overt signal is the only possible signal?[63] Many women report that they are aware of their ovulation and that they respond differently sexually during midcycle (See Chapter Three for a more detailed discussion of this issue.) We are a species covered in clothing, doused in deodorant, and washed clean with shampoo and soap. Stripped of those cultural perfumes and disguises, as Lucy was, we too might signal our ovulation like a chimpanzee in heat.

THE LURE OF THE BREAST

Stand in front of a magazine counter sometime and count the number of breasts. They seem to be everywhere—on covers, in advertisements, as feature stories on breast enhancement or reduction. The female breast is big business in our culture. Other mammals—the family cat or dog or animals at the zoo—also have the same sort of tissue that excretes milk for nursing babies. But only humans have evolved soft fatty cushions for their nipples. Anthropologists speculate that breasts, in the modern human form, have figured prominently in our sexual evolution.

Most anthropologists feel that fatty and pendulous human breasts evolved as a sexual signal to males.[64] At the most basic level a good-sized breast signals fertility; it shows that a female is reaching reproductive age. This makes some sense, since young girls grow breasts at puberty as their reproductive organs mature and conception becomes a possibility. Protuberant breasts are therefore straightforward signs of their impending fertility. Perhaps, some also suggest, fatty breasts indicate a female's nutritional status, which in turn would signal to a male that she could sustain a pregnancy and lactation.[65] However, breast size, contrary to popular belief, is not correlated to milk production; size alone is not necessarily an honest signal of female nursing ability. Others suggest males should be looking for symmetrical breasts, breasts with good nipples capable of prolonged nursing.[66] Human milk is low in protein, fat, and calories and our species is adapted to infants feeding on demand throughout the day; good nursing ability would be important.[67] The only problem is that there is no clear relationship between nipples, breasts, their shape or size, and the ability to nurse an infant efficiently. The point is, even if men were interested in choosing women with the best breasts designed for extended nursing, there are no clues that would necessarily indicate what makes a "good" breast. The evolution of female pendulous fatty breasts, then, can't be traced to what men might be interested in.

It's also possible that fatty breasts have nothing to do with communicating to males. A large breast might be simply a fat storage area for females who evolved under nutritional stress.[68] Ancestral humans walked long and far in their search for food, and they needed fat storage for years of lactation. The breasts, and the rest of the female body for that matter, might be designed to help store fat. Fatter women always give birth to fatter babies, and infant size is certainly important to survival in a harsh world.[69] Fat deposits are found in breasts and on the hips of human females and give them that distinctly human female shape, a shape which coincidentally makes carrying large babies a bit easier.[70] From this perspective, the fatty human breast is less a sexual signal to males than a helpful mechanism for improving a female's chance of keeping a child alive and healthy. Although our American and European culture has highlighted breasts as something sexual, the reaction on the part of males may be more culturally molded than based on a biological urge. Fatty and pendulous breasts might, in fact, be more important to women than men, at least from an evolutionary perspective.

THE MALE ACCOUTERMENT

Most male mammals have a bone in their penis called the bacculum. This bone presumably evolved to aid in the deposition of sperm deep into a vaginal canal. Human males, for some reason, lost the bacculum at some point in their evolutionary history. But they gained a rather large penis, about 13 centimeters in length on average, that outdistances those of both chimpanzees and gorillas. Despite the size of the human penis, the testes are relatively small—larger than those of gorillas, but only a fraction of the size of chimpanzee testicles. There's no correlation between penis size and sperm production; there is, however, a relationship between testes size and sperm production. Humans, for example, make less sperm than chimpanzees, but more than gorillas.[71] Why then do human males have such large penises?

One possibility is that a big penis can frighten away other human males.[72] Men spend a great deal of time competing with other males, sometimes for access to females, and the large penis may have taken on the role of display, a measure of a man's aggressive prowess. This idea is supported to a degree by the fact that in all cultures today men always cover their penises, even if they wear no other clothing. Large penile size only comes about at puberty, when the penis grows exponentially. It is at this point in male maturity that male-male competition for mates becomes important. Perhaps the human penis acts as an implicit threat, somewhat like the long canines of monkeys and apes.

But there's another possible hypothesis that explains the evolution of a large penis. This hypothesis suggests that human females, concerned with the quality of their sexual encounters, choose as sexual partners men with large and satisfying penises more often than other males.[73] As a result, large penises have been selected over time in our species. Unfortunately, this intriguing hypothesis has never been tested. At present, we still don't know why human males stand out, in terms of their penis size, from their ape relatives.

A Scenario for the Human Sexual Animal

The natural history of our sexuality can't be separated from the rest of our human biology. We are large bipedal mammals who give birth to highly dependent infants that grow up to be dependent juveniles. Our reproductive success, that is, how we pass on genes, has evolved over millions of years and is focused on these intensely needy packets of genes we call children. The history of the primate order is one of babies becoming more and more altricial and needing caretakers, and although we might not be able to explain why this form of human animal has evolved, it must be a reasonable strategy to have survived so far. The pieces of the puzzle suggest that our ancestors, perhaps as late as 2 million years ago, lived in small groups of unrelated females and several males who might have been related.[74] These groups moved about the savanna, stepping in and out of patches of forest, gathering vegetable matter as they went. They probably didn't defend any particular area since their large body size required large tracts of land to sustain them.[75] At some point, these early hominids, or humans, began to exploit meat as a food resource. They probably started as scavengers, lurking behind lions and hyenas, waiting in the noonday sun for everyone else to fall asleep so they could sneak up and drag off a carcass. They were already bipedal so they could easily carry meat to a safe haven. With large brains, they became excellent strategists and used their wits to stay alive and get enough to eat. They may also have used puzzle-solving skills to wipe out others who shared their resources and land. And someday down the line they became organized hunters, artists, and computer programmers. But what of the relationships between group members?

Humans are no different from other social animals. They develop attachments, choose mates, have sex, and produce babies. Although we might like to believe we come from finer stock, our ancestors, just like chimpanzees and gorillas, had sex because their genes pushed them to do so. And our ancient human sexual system was molded by the need to obtain enough food to live and protect oneself from predators. The clues to our sexual biology include a difference in male and female body size which was dramatic 3 million years ago, but decreased over generations. This dimorphism points to a polygynous heritage with males mating with several females and females having access to several males. Although some might think we've moved toward a more monogamous way of life to accommodate dependent babies, it's also possible that we became more pair-bonded as parents but have retained our less than monogamous sexual freedom, even to this day. The lack of sexual swellings on the backsides of human females shows that probably today, and in our past, hominid females did not need to advertise their fertility nor their sexual readiness. But human females also have a long history of sexual flexibility. Human males, like all male animals, have the potential to have sex quite frequently, but compared to other animals, this potential is actually rather low (see Chapter Four for further discussion).[76] Men are hindered by low sperm counts and long refractory periods in which to manufacture more sperm, and by access to a large group of interested females.

Both sexes have the potential to be highly sexual beings. Our sexuality is not seasonal or limited by estrus. Because females and males can have sex whenever the urge strikes them, sex is most often, in our particular species, disconnected from reproduction. In fact, until the last two hundred years or so, the human population had a rather slow rate of growth.[77] We have sex but much of that sex is nonconceptive. Humans start reproducing rather late in life; we have a low rate of conception each cycle; we have long interbirth intervals between each child; we have a history of using birth control and consciously trying not to conceive. And in this we are somewhat different from other animals, for whom sex is most often the expedient route to passing on genes.

Our ability to have sex without the consequences of reproduction is the obvious fact behind the red light district of the Oudeizijdsachter canal and the back streets of Bangkok. Sex evolved as a pleasurable response to ensure that each of us is pushed to mate and pass on genes. But that incentive has been extended and become a major feature of our species. We are sexual beings, just as all animals are, but we tend to exaggerate our sexuality, as with so many of our behaviors. At this point in our evolution, sexuality defines so much of our culture and our lives. In the next three chapters, I explore the biology behind our sexuality that evolved over many millions of years. I first ask the most fundamental of questions—what makes us have sex?

CHAPTER TWO

The Essential Urge

A friend once told me that the average person has a sexual thought every fifteen seconds. This is one of those unsubstantiated factoids that somehow sticks in the mind. Sometimes when I'm teaching my large anthropology class, this pronouncement sneaks its way into my consciousness right in the middle of a lecture on, say, early human tool use. I suddenly wonder exactly what my students are thinking about. While I try to explain why a bifaced hand-ax is an important step in our evolutionary history, I imagine these eighteen-year-olds afloat on hundreds of sexual images. Are they remembering some moment of ecstasy from last Saturday night, or maybe sizing up the person in the next seat for next weekend? And in that instant, when I ponder these questions, aren't I, too, having a sexual thought, spending precious time thinking about the sexual thoughts of my class? Then I realize that

I'm actually doing two things at once—thinking about matters sexual *and* lecturing.

Although our sexual self is embedded in our daily lives, it pops up at the oddest times, seemingly out of nowhere. We go about our day, dealing with everyday tasks, attending to a million little decisions—what to cook for dinner, determining why the baby is crying, wondering whether or not a presentation will fly at work, and thinking up fun activities for the weekend. Woven in among those thoughts and actions is our sexuality. Some of us think about sex frequently and act on those thoughts, while others go hours, days, or months seemingly without a sexual impulse. But no matter the time interval, no matter the intensity of our sexual images or fantasies, sexuality is an issue for every individual, an issue that at times consumes our thoughts, fantasies, memories, involves planning and preparation, and fills us with hope, disappointment, happiness, or even despair.

The Tease

What is this strange feeling that bubbles up from nowhere? Why can't we, some of us sometimes wonder, just go through life ignoring our sexual side and get down to the business of living? We can't because the drive to have sex is part of our inborn nature. It's helpful to compare the sexual urge to another of our basic human instincts, the desire for food. One moment you may be blithely walking down the street when suddenly you feel an ache in your stomach, a signal to your brain that it's time to eat. Up ahead is a sandwich shop, and you stop in and pick up a turkey on rye. But the signal that prompted the sandwich purchase is more complex than that. Your stomach is actually telling you that your blood sugar is down, probably because it's been several hours since you took in enough fuel to bring it up to satisfying levels. For some unknown reason, your mind craves a turkey on rye today, rather than a slice of pizza. Although we might think we understand the overall cycle of hunger-food-craving-satisfaction, the average person usually doesn't think about or understand each separate element of the cycle. Nor can scientists explain why one day an individual reaches for a turkey sandwich and the next day craves a doughnut. Most experts suggest that food satisfaction is a mixture of chemical changes inside the body, psychological triggers, and basic biology. In that sense, sex is a similar kind of impulse; it too involves chemistry, psychology, and basic biology. These several factors, when combined, explain why we have the urge to have sex. While none of them alone is responsible for sexual arousal, together they signal the motivation to follow through with that sexual drive.

MY GENES MADE ME DO IT

Almost daily, the newspapers seem to announce another connection between a particular gene and a certain disease or behavior. For example, depression, hemophilia, myotonic dystrophy, Tay-Sachs disease, and many other conditions have been mapped to specific genes on particular chromosomes. Within the next fifteen years, the Human Genome Project is supposed to draft out the cartography of our DNA and reveal the place on our chromosomes of many other diseases and hereditary traits. There's also recently been a flurry of research into a possible link between homosexuality and genes. By implication, if homosexuality has a genetic component, so too must heterosexuality, or any kind of sexuality. And if that's true, genetic research has the potential to unlock more of the secrets of our sexuality. But of course neither humans nor any other animals are genetic automatons.

THE GENETIC BASIS OF BEHAVIOR

Although the genetic map may soon be able to pinpoint certain malfunctions in our DNA, it won't be able to explain most of our behavior. This is because we—and all animals and plants for that matter—are a complex mixture of genes and environment. The monkeys I study, macaques, run their lives through a maze of hierarchical relationships. High-status animals have better access to food, mates, and shelter. They also have social power; lower-ranking monkeys want to sit close to them, or elicit their help during fights. Low-rank animals are often more peripheral, less healthy, and they don't produce as many offspring. Young females stay forever in their natal groups while young males leave at sexual maturity and transfer into neighboring groups. Primatologists have noted that a daughter's rank is like her mother's; if the mother is of high rank, so is the daughter, and if the mother is of low rank, her daughters are too. In other words, daughter macaques generally "inherit" their ranks from their mothers. This does not mean that there is a place on a macaque chromosome that determines rank. Instead, daughters learn by watching where their mothers fit in, and they act likewise. At the same time, there is also a real genetic component to social position. Size, health, and certain abilities are encoded on genes, and these too add to how a monkey acts and which position she might achieve. When young males leave the troop they don't have mothers to help them figure out a place in their new community. Instead, their body size, fighting ability, and the behaviors they learned as youngsters all determine where a male ends up in an adult male hierarchy. Because genes and environment are involved in status position, the community's hierarchy is a dynamic interaction of monkeys. A female can change her position by fighting and moving up in the hierarchy, or she can fall in rank for various reasons. A male can never predict his status from one year to the next. There might be genes which contribute to rank, but it is the interaction of genes and environment that models the system.

The same scenario applies to human behavior. There may be a genetic component to our behavior, but even if there are genes involved, environment also has an influence. Our sexuality may have some genetic basis, but how we grow up, who we are, and the culture we live in are just as important in molding our sexual self.

The second reason there's no easy step from genes to behavior is that there's no one-to-one correspondence between this or that bit of DNA and a specific behavior. It's relatively easy to come up with the genetic recipe for something like insulin or the hormone Human Growth Factor because they're straightforward chemical combinations of DNA. But it's impossible to pinpoint the genes for a pattern of behavior because behaviors are a series of decisions and events, not a simple chemical reaction. What, for example, is the DNA sequence for a complex behavior such as laughing? I can state categorically that no single gene for laughter exists. But does that mean our genetic makeup has nothing to do with how and why we laugh? No. Genes have a great deal to do with it. We have genes that allow us to assimilate the words spoken as a joke, react with a smile, and control our breathing in silly-sounding puffs that expel air. The only animals that regularly laugh at something funny besides humans are chimpanzees, and chimpanzees have many of the same genes we do.

More to the point, the question of a single gene matching to a single behavior is spurious. In the same way, no single gene fuels sexual behavior. Scientists may, at some point in time, clearly link our sexual orientation with a particular point on the gene map,[1] but so far you can't locate exactly where on which chromosome is a spot that motivates us to have sex. But just because we can't point to a particular piece of DNA on a particular chromosome with regard to sexual behavior doesn't mean that genes have nothing to do with our sexual behavior. Quite the contrary. Genes, in fact, have everything to do with why we have sex; they're the biological spark plugs that motivate us to seek out and couple with another person.

CHOOSING SEX

No one likes to hear that something other than their own free will is guiding their behavior or their life. This is especially true in Western society, where free will is a basic tenet of our political and economic system. It wouldn't surprise me if some readers experienced a knee jerk mental protest to my suggestion that our genes underlie our urge for sex. But from a biologist's point of view, the basic reason, the fundamental reason, we have sex is because our genes tell us to.[2] If it sounds like I'm suggesting we're behavioral automatons, brainlessly moving from body to body, spreading our legs or spreading our sperm, that isn't really what I mean. But think for a moment about sex in another animal species, such as a lizard. If I point out that lizards copulate because their genes are encoded to make sure every lizard mates and reproduces itself, no one would blink. This is because watching two lizards mate doesn't exactly conjure up a romantic image of shared lizard pleasure, or of lizards moving to cement a special relationship. We see instead Nature at work. I could probably convince you with relative ease that those lizards are, in a biological sense, two bundles of DNA joining up so that they have the opportunity to replicate their DNA and pass it on. But when I suggest that the same type of human-encoded DNA is behind all the craziness we call love in our species, people bristle. We want to believe human copulation is something more lofty, more charged with meaning, more of a mind/body interaction than it is with other animals. This may often be true. But way down deep, underneath all the love and lust, underlying the attachment and intimacy, is DNA pushing us along.

REPRODUCTION THE SEXUAL WAY

From the first time a one-celled organism moved to join its DNA with another one cell, the sexes have been combining to pass on genes. We humans are no different from lizards or monkeys in this. We're compelled to have sex because all sexually reproducing creatures before us have passed on genes in this way. Oddly enough, no one is really sure why or how sexual reproduction began in the first place.[3] Long ago all organisms produced by cloning, that is, duplicating themselves like a Xerox machine, dividing into sister cells which are genetically identical, and passing on exact copies of themselves to future generations. At some point, about 3 billion years ago, some one-celled bits of life reached out and combined. Perhaps they needed foreign DNA because theirs had been damaged by sunlight, or perhaps they were low on nutrients and water and sought other cells as a prey; in any case, the original cell now incorporated new DNA by gobbling up another. This was the ancestral version of sexual reproduction.[4] The refashioned cell was not exactly like either of the originals; it had a genetic blueprint never seen before because it was a combination of the two "parents." But why was that advantageous to survival and why did sexual reproduction continue and

38

almost impossible to tell the gender of most birds, and Einstein got his name under the assumption that he was male. But one week Einstein began laying eggs. We hovered in amazement around her cage, picking up the lovely pink-white orbs that she dropped over the course of a week. They weren't, of course, fertilized, because Einstein lives alone, but her eggs were a dramatic demonstration that reproduction proceeds even if sex is absent. Urban pet owners are aware that if they don't do something to curb the natural urges of their pets, they will be involved in animal sexual exploits. We neuter male cats because males roam if some-one doesn't disconnect their testes from their brains, and we fix our female cats so that they won't be compelled to writhe and moan in an attempt to attract suitors. What we see in the animal world around us is sexual behavior that seems automatic. When a female cat goes into heat, she seeks males. When a male dog smells a receptive female, he tries to mount her. Einstein lays eggs even without a partner. Yet we imagine that humans are less "animal" in this arena; we believe high intelligence must be an interface between the power of genes and sexual behavior.

As a result, humans set themselves apart from other animals, especially in the realm of sex. We somehow believe that our species is less biologi-cally compelled than other animals to engage in sex. Instead, humans often cloak sexuality in romance and love to disguise the basic reality of reproduction. While the act is often less immediate, and we usually do engage in at least a minimal amount of decision-making before we have sex, we too are compelled by our biology. Yes, we make decisions about sex, and about the partners we choose to have sex with, but just like our pets, the fundamental urge to have sex, based in our genes to encourage us to pass on copies of our DNA, is part of our makeup, as much as the sexual whisperings of any animal. We may have more neurons in our brains than a parakeet that has no conscious thought about sex, and yet hoists its rear into the air to receive sperm from a male when in heat, and we may not call to males across the fence or sniff out females at the local hydrant, but still, the need for sexual coupling is basically the same. If we didn't have this preprogrammed urge to mate and pass on genes, we'd never have sex at all.[7] "Good" genes, from an evolutionary point of view, are those that make it to future generations; they have been en-coded to ensure this. Genes that are less successful in making sure the individual they reside in passes them on are, by definition, wiped out. And so it's a self-defining process. Genes that make sure an animal mates have a vehicle to pass along their genes. And so those genes make it, while the other genes, those that let the breeding season pass, or don't motivate an individual to find the right partner, lose out. This is how evolution works at the level of sexual behavior. But while our genes might explain our urge to procreate, do they explain our ever present obsession with sex?

THE PLEASURE OF SEX

Apart from genes, why would anyone have sex? Well, because it feels good, someone might answer. I would counter with the question "But why does it feel good?" For some reason, our minds and bodies connect physical pleasure and satisfaction with sexual coupling. This connection, too, is tied to evolution. Sexual arousal is a specific feeling that tingles the genitals and begs for further stimulation and cries out for eventual release. The point is, the pleasure factor is linked up to sex because we do indeed have to be motivated to have sex so that reproduction can occur. Sexual pleasure is simply a physical reaction that evolved over millions of years to make the genitals sensitive and the body able to orgasm. Without the element of pleasure, we wouldn't be driven so powerfully to have sex. And if someone doesn't bother, his or her genes aren't passed on but are eliminated from the next generation. In other words, pleasure as a motivator for sexual intercourse is part of the very same human DNA that ensures we copulate.

Let's compare the sexual instinct with another human instinct, the need for food. We love to eat. In fact, we often eat more than our bodies need just to stay alive. Most of what we eat tastes good, and given the option, we choose things to eat that make our mouths happy. Our ancestral genes have encoded our bodies to mentally react to the taste of food by recognizing a certain kind of stomach and mouth satisfaction which in turn encourages us to eat. If we didn't like the taste of food, if it didn't give us such enormous physical and psychological satisfaction, we probably wouldn't be as inclined to eat, and we'd potentially die of starvation.

Evolution has given us a sense of taste which encourages us to eat a variety of foods, and the end point of keeping our bodies alive has been served by the means of tasty food.

In the same way, our genes have made sex pleasurable to serve the end of passing on genes. These genes are not conscious, there is no morality involved. Genes are simply designed to replicate and be passed on. If the individual decides not to reproduce, or is unable to for any reason, those particular genes are lost to future generations. Their loss is only evolutionarily important because any species is made up of the collective of its gene pool, and those lost genes will not add their particular character to the pool.

The Biology of Lust

It's strange how the feeling can strike anyplace, anytime. There you are sitting in a café, sipping an espresso, reading the morning paper, and suddenly you feel a surge of sexual desire. This inner voice comes seemingly from nowhere—there's nothing in the paper the least bit erotic, no cute woman or sexy guy at the next table, and you may be only marginally awake anyhow. One reason our sexuality is such a mystery is that our urge for sex pops up at the oddest times. It's easy to understand why watching an erotic movie would arouse, or the sight of a particularly attractive potential mate leaning against a bar in tight pants, but why does the urge often materialize seemingly out of nowhere? Like all other human impulses, such as hunger or the need for sleep, it doesn't really appear out of nowhere. There's a source for that feeling, but it's one that is just a little harder to pin down.

THE SEX DRIVE

Everyone has a sex drive, or libido. But as with all other animals, there's a large variation among people in how much a person is motivated to actually act on that libido.[8] The media might like us to believe that everyone is a sexual dynamo, constantly fixated on matters sexual, but this isn't true. Just as in all other appetites, for say chocolate or knowledge, we differ in how much we're interested in sex. Each one of us has a particular "sexual arousability" or capacity to respond to erotic stimulation, and those with a high level of arousability will be more motivated to seek out sexual interactions. No one knows yet *why* some people have higher sex drives than others. It might be a biological difference, something prewired from birth, or a difference fostered by various upbringings. Surely each one of us has experienced conflicts in this area. There's nothing more frustrating than being paired up with a partner who has a different sex drive. Any sensitive person caught in this dilemma might assume, when they are the one consistently rebuffed, that the loving partner just doesn't find them attractive. Or the less interested partner may label the other as "weird," "sick," or "oversexed." When I think about the difficulties inherent in two sexually mismatched people, I'm reminded of that scene in Woody Allen's movie *Annie Hall* where Annie and Woody are each talking with their therapist and we see their sessions on a split screen. Annie says in exasperation, "We have sex *all the time,* three times a week." And Woody says with the same frustration to his therapist, "We *never* have sex, only about three times a week." The point is that each of us has a certain motivation to have sex. There might be an average, but this isn't the norm because there *is* no norm. The human sex drive is a continuum from "never interested" to "always interested," and all of us fall somewhere along that continuum. Our specific sex drive is influenced by an internal biology and psychology, and all sorts of other intrusive layers of our lives. No one is surprised when I suggest that there's variation in every human ability, such as intelligence, artistic sense, appetite, emotional sensitivity, or physical strength. Sex drive is another behavior pattern that is just as variable.

THE LOGIC OF THE LIBIDO

Although we might all be individuals when it comes to the degree of sex drive, we do share the basic mechanics of that libido. Psychiatrist and sex researcher John Bancroft sees our sex drive as an interaction between what goes on internally in our bodies and various stimuli from the out-side.[9] Like me, he likens it to an appetite for food. Internally, we all experience the biochemical and physical changes that tell us to eat. Our blood sugar goes down and our stomach grumbles when there hasn't been some caloric input for a while. At the same time, external stimuli such as the odor of baking chocolate chip cookies or a picture of greasy French fries on TV can speed up an already existing hunger, or produce a new sense of hunger in a person with a full stomach; obviously these external factors, too, influence appetite: they make us salivate for fat and sugar. At the same time, both one's internal drive for a salad or an external push for candy, such as a Snickers bar advertisement, can be sidetracked by psychological forces—concerns that you'll eventually look like a baby hippo if you give in to your cravings for snack food, for example. Bancroft sees the sex drive in the same way. The internal force is a basic instinct to have sex, the inner voice that desires sex as part of a healthy life pushed by genes that compel you to procreate. External sexual forces in our culture are many—just about any sight, odor, or sound, when coupled or associated with the right sexual image, can be considered sexual and can pique our basic sexual instinct. But the sex drive can also be easily derailed by the psychological—ghosts from our past, interruptions, or a tiring day at work. In that way, the sex drive is just like the appetite for food. Both are first and foremost a basic instinct, one to nourish and maintain our bodies, and the other to pass on genes. Both can be influenced by outside forces such as sights and odors, and both can be sidetracked or further stimulated by psychological issues.

Given that we all have a particular sex drive, an instinct of sorts that can be kicked into action, how exactly does this work? The coffee shop scene depicted earlier is a good starting place because its role in sexual arousal is not as innocent as it first appeared. Our male subject was sitting quietly minding his own nonsexual business when a little inner voice woke up to sexual desire. This voice didn't really come out of nowhere. The man may have actually been stimulated into thinking sex-ual thoughts by the simple smell of coffee. Let's say that the last time he had sex was the previous Friday night. He'd gone to the movies with his girlfriend and went out for dessert later. They both consumed pieces of death-by-chocolate cake accompanied by espresso. Then they went home and made passionate love. The coffee smell and thoughts of sex were melded together and filed away into his unconscious, only to be reawakened by the smell of this morning's espresso. This is not to say that the association between coffee and sex is now a permanent pathway in that man's brain. It isn't. But at least for this moment, on this morning, the connection triggers a sexual spark.

In all our sexual impulses, we are always first triggered by a stimulus. The difficult thing for researchers to predict is exactly what that stimulus might be. And this is a general problem for all us mammals. As brain size increased over millions of years, we mammals gained a lot of mental maneuverability both in the sexual arena and in all thought pathways.[10] While increased brain size and complexity allows us to think though complex puzzles and solve ecological and social problems, it also means we're a little less "hardwired" in our thought processes than other creatures who don't have to process every input before reacting. In other words, a bigger brain can be a burden, too. In terms of sexuality, the roots of our sexual motivations and reactions are somewhat more confusing than other animals' because of this large filter, the brain, through which all our sexual impulses pass. In snakes, for example, it's pretty easy to draw a straight line between a certain stimulus and the move toward coupling. The male snake with its tiny brain, for example, sees a female in the correct mounting posture and his brain says, in essence, "Mount her." But for large-brained mammals, it's often impossible to tell exactly why something arouses us. And since humans have the biggest brains of all, there are millions of neurons that fire in response to all sorts of input that triggers our sexuality. Alfred Kinsey, the famous sex researcher who interviewed thousands of Americans in the 1950s about their sexual behavior, said that nearly everything in the world is a sexual stimulus to somebody.[11] Some people are stimulated by the smell of leather, or a bouquet of roses, while for others, the smell of coffee fires up their neurons. And what might be a mundane word, smell, or picture one day can turn into erotica the next. The sexual association of all these stimuli is based on our previous experience and conditioning, and by what our culture defines as sexual. For example, bare breasts in America are exciting, but in Bali the simple sight of breasts exposed to the public doesn't mean much, and the sight of female genitalia is a sexual stimulus worldwide.[12]

SEXUAL SENSES

Although humans are vulnerable to any number of sexual incentives, some forms of stimuli are more powerful than others. Humans, it seems, are more affected by the sense of touch than by any other type of stimuli. Sight or smell might trigger sexual arousal, but that urge is usually translated first and foremost into a compelling need to be touched. Touch doesn't have to be mediated by the higher centers of the brain and can therefore trigger a reflex sexual response.[13] And certain areas of our bodies have evolved especially to be stimulated in a sexual way. These erogenous zones, the lips, buttocks, genitals, and such, are geared to be the nerve endings of sexual experience. Once these areas are touched, the sexual impulse speeds up. This hypersensitivity to the sense of touch is also seen in other areas of our humanness. As social beings, humans are often in contact with other humans. Unlike most other mammals, humans and most other primates live in groups and spend hours in contact with their own kind. Our need for touch is also reflected in our anatomy. We have evolved away from claws and more toward fingers with nails and fleshy, highly sensitive pads on the undersides of our fingers that act like physical radar.[14] We like to hold hands, sit close, and touch our friends in passing. We even shake hands to make a deal or say hello and goodbye. Touch is used every day by people to signal intimacy because it's a physical sensation that the neurons of our brains take from the fingertips and register as closeness. It makes sense, then, that our sensitive fingers would be sexually stimulated when they touch the skin of another, and that our other body parts would respond when touched in turn by those soft-bellied fingertips reaching out to make contact.

Humans are also sexually triggered by visual stimuli. This, too, makes sense in light of our basic nature. We are primates and have evolved from a life in the trees.[15] One sign of our primateness is a reliance on vision, including color vision and good depth perception. Our ancient ancestors needed to pick out ripe fruit and be visually able to track their troop. They also needed to jump and swing through the trees, sometimes racing faster than predators on the ground. As a result, our sense of vision is keen.[16] Our eyes face forward where they receive images and send them back to both sides of the brain. Primates' eyes are also set well into our skulls, protected by a socket of bone, signaling their high value. We may not see as well as eagles and falcons when they spot a field mouse from a treetop, but we do rely on vision as a means of interpreting the world. A mouse presented with a photograph of copulating mice wouldn't react much because it doesn't have the visual ability to recognize that flat piece of paper as an image of mice mating, and his brain can't process the information anyway. In contrast, humans are easily aroused through the visual channel. Sex scenes in movies are placed there because we are victims of our tree-living heritage that gave us great visual acuity, as well as a big brain to interpret and empathize with the writhing image of two movie stars in bed.

The sense of smell, our male subject's coffee attachment aside, is less likely than touch or sight to activate the sex centers of the human brain. In our evolutionary history, we lost the sense of smell but gained an increase in visual acuity. In particular, we lost a system called the vermonasal that communicates a reaction to odors directly from the intake site, the nose, to the limbic system of the brain.[17] As a result, primates have a poorly developed sense of smell. Look at your own face— the lack of a decent snout should tell you something. Compare your face to that of the family dog. Now there's an animal that lives and dies by the sense of smell. It's hard for a human, or any primate, to imagine the world of a dog because that world is so dominated by odors. They smell food, people, and other dogs, and the more putrid an odor the better. The sense of smell is also part of why a male dog is attracted to a female dog in heat. For humans, an odor has to be pretty strong and keyed to a particular memory to play a role as a sexual stimulus. But that doesn't mean smell doesn't matter at all. For some men and women, perfume can be a sexual trigger; certainly perfume manufacturers hope there's a bit of dog in each of us. We excrete odors from our sweat, urine, feces, breath, saliva, and skin.[18] Women also produce volatile fatty acids in the vagina which change over the cycle and seem to affect the cycles of other women (see Chapter Three). In one classic study, subjects were asked to wear a T-shirt for twenty-four hours with no deodorant.[19] Afterward, each subject was presented with alternatives—their own T-shirt, a T-shirt of a strange male, and the T-shirt of a strange female. Eighty-one percent of the sixteen males picked out their own T-shirt, but only 69 percent of the thirteen females were successful in detecting their own odor. All the subjects were pretty good at discriminating between male and female shirts. Clearly, odor doesn't play as much of a role in our sexual lives as it does in other animals', but this doesn't mean smell has no effect at all on sexual arousal and choosing partners.

Sound, like smell, is another sense that is not particularly important as a sexual stimulus for humans. Sure, certain music is "sexy," but just playing that music won't have much of a sexual effect without touch and sight. And some voices, over the telephone for example, can be stimulating. But still, sound is not as important to primates as the sense of touch.

Although touch and sight are the key avenues of sexual stimulation for humans, one signature of our species, and perhaps one curse, is our ability to be sexually aroused by just about anything.[20] While most stimuli are common to all of us—naked people, smells of those we lust for, a memory of great sex—others are more particular to a minority. Most of us think of rubber as something that makes nice tires, and bare feet as appendages that cry out for cowboy boots, but others become sexually aroused by these objects. The complexities of human psychology explain the development of some of these sexualized objects, but the point to note here is that one mark of our human sexual nature is the potential sexualization of just about anything our minds adhere to.[21] Once the stimulation has been introduced, however, our bodies react in pretty much the same way. The mechanics of sexual arousal is one human universal that differs little from male to female, African to European, heterosexual to homosexual.

Body Heat

Overall, sexual experience is characterized by changes in the genitals and other parts of the body, a sense of heightened awareness, and specific neurological interactions between the brain and the rest of the body.[22] Although we experience sex as a purely physical sensation, it is the brain that interprets and filters those sensations. But much of how our bodies react is automatic. Just as breathing, elimination, and metabolism are involuntary systems that we rarely think about unless something goes wrong, sex, too, has its automatic responses. The physiological or mechanical process of sexual response follows the path of any automatic physical reaction. During initial arousal, a person is usually aware to some degree of how his or her body is reacting and they can stop or continue at will, even though the physiology involved is purely automatic.[23] Perhaps this is why many people see our sexuality as something slightly animal-like, because it's one of those bodily functions that we control less tightly than others. Think of another involuntary automatic system—breathing. We go on breathing every few seconds, awake or asleep, day after day. At the same time, we can consciously control breathing, as when we are talking, singing, or shouting. Sex too is partially an automatic physiological response system which is turned on and essentially extended by the mind.

BRAIN STORM

No one is exactly sure which areas of the human brain are involved in sexuality. It's much easier to track the route from sexual stimulus to brain to physical action in life-forms with smaller and less complex brains than humans. But even then, it appears that there are a number of sites in the brain that respond to sexual stimuli, and these areas aren't necessarily situated next to each other.[24] In addition, particular areas seem to control separate aspects of sexuality. For example, when a lesion is cut into the medial preoptic-anterior hypothalamic region in the male rhesus monkey brain it will impair his ability to copulate but not his ability to masturbate.[25] In the only study of female primate sexual behavior and the brain, researchers demonstrated that a lesion in the brain of a female marmoset canceled her proceptive behavior, that is her sexual assertiveness, but she was still receptive to the approaches by males.[26] Scientists also know that rats have an area in the forebrain, called the medial forebrain bundle, which is the center of rat pleasure. If you give a rat a choice between stimulating that forebrain or eating, he'll choose the pleasure center every time.[27]

For humans, brain tissue involvement in sexual arousal is a bit more complex. Parts of the hypothalamus, an area deep in the center of the brain that's important to reproduction and body rhythms, are involved; in addition, there may be several hot spots throughout the largest part of our brain, the dome of the cerebellum.[28] But no one has been able to pinpoint the centers of sexuality. In fact, stimulation of parts of the neocortex during brain surgery have never caused anything that could be construed as a sexual response from a patient.[29]

Sexuality also registers in the most primitive part of our neural system, the brain stem, and down the spinal column. For example, the nerves that control an erection are low down on the spine, essentially opposite the penis. John Bancroft describes this response as a cognitive or thinking process which influences the limbic system, which in turn affects the spinal cord and causes reflex reactions through nerve pathways to the genital areas and other body systems.[30] Once the process has begun, the mind is eventually conscious of the process, which completes a cycle. In a sense, the sex act is psychosomatic since it involves both cognition, or thought, and the physical responses of the body.

Once a sexual stimulus—whatever that stimulus might be—triggers nerve endings, impulses are sent to the brain. Neurons responsive to sexual signals are then alerted, and react by sending impulses to various parts of the body. It's a loop of stimulus, impulses traveling from the stimulus point to the brain, and back out again to other parts of the body. Bancroft compares the nervous system's role in sexual arousal to the electrical wiring of a house.[31] Most appliances and all the lights work on a generalized wiring plan that contain a multitude of on/off switches. But if a heavy-duty air conditioner is anticipated, the electrician might string a special wire from the fuse box directly to the air conditioner. This wire bypasses all the other bells and whistles and will turn on only with its own special switch. Sexual arousal is like the wired house in that stimuli will send a current through all parts of the body to the brain, where it is interpreted and causes a reaction. As a result, the heart rate increases and breathing becomes more rapid. The same stimuli cause an immediate and direct reaction in other parts of the body more directly wired to their automatic response outlets, much like the air conditioner wired to its special switch. In the case of sexual arousal, the wires are directly connected to the genitalia. Once sexual arousal has begun, the penis and the vagina are among the first body parts to respond to the stimulus. This reaction is in part mediated by the brain, but it's mostly driven by a direct physiological path, a direct wiring, that isn't really "willed." It's as if someone walked into the house, turned on all the lights by flicking on every switch, making the electric current zip in and out of every connection, and at the same time, the home owner turned on the special switch to the air conditioner. The air conditioner has a direct wire to the source of the electricity and so it doesn't wait for the current to run through the rest of the house before it starts to work. In the same way, the vagina and the penis respond to sexual stimuli even before the heart rate has gone up. On the other hand, the responses of the vagina and penis, and the rest of the body, aren't so automatic that the brain can't jump in at any second and shut the whole system down. In other words, the automatic sexual system doesn't mean an individual is out of control, it just means that not a whole lot of thought is needed to flick the switch of sexual arousal.

THE SIGNS OF AROUSAL

Perhaps the most obvious sign of sexual arousal in our and other cultures is an erect penis. Women can fake arousal, but a man without an erection is obviously not aroused. We might conclude, visual primates that we are, that men alone have a strong physiological response to sexual signals. This, however, is not true. Given the appropriate sexual signals, the genitals of men and women react exactly the same. Although women in Western cultures always protest they're not as sexually motivated as men, experimental data show that women and men are in fact just as easily aroused.[32] In one study, the similarity in arousal between the sexes was documented by measuring changes in male and female genitals while the subjects watched sexually explicit films.[33] Male subjects wore a penile strain gauge which measures how fast and how much a penis swells. Woman were fitted with a photoplethysmograph, a tampon-like device that fits into the vagina and measures vaginal temperature and blood flow through vaginal tissue. The photoplethysmograph is a light-emitting diode encased in a tampon-like tube. Once inserted, it shines light into the vagina; the amount of light reflected back is relative to the amount of blood in the vaginal wall tissue. When a woman is aroused, the blood rushing to the vagina will absorb much of the light, but when she is relaxed, the light shines back. The degree of reflection, which is transmitted to a polygraph, measures the pulse of blood through the vaginal walls and the pooling of blood in the tissues.[34] The subjects were exposed alternately to erotic films and nondescript commercials intended not to arouse them. In between the commercials they also did a few math problems to clear their heads of sexual thoughts before they were shown another bit of erotic film. Control groups were shown the same commercials, but the films they viewed were about lap quilting, a neutral stimulus, to make sure the people being tested were not aroused simply by wearing the measuring devices and watching TV. The researchers found that early response patterns to the erotic films, according to the physiological measures of genital arousal, were the same for men and women.

Physically, sexual arousal is initially experienced by both sexes through vasocongestion—increased blood flow—and myotonia—muscle tension —in the genitals. The muscle action, of course, affects vasocongestion since the contraction of muscle tissue inhibits or releases blood flow. During sex, the parasympathetic nervous system, a part of the autonomic nervous systems that acts automatically, controls the contraction and relaxation of smooth muscle which directs blood into certain tissues.[35] For men it works like this: At the base of the spine, what we more often think of as the lower back, is a group of neurons called the "erection center." I can only imagine an Erector set with cranes and pulleys ready to receive the signals of a few nerve impulses. Once these neurons are aroused, they send a message to the penis to let the arteries open slightly and allow excessive blood to flow into the penile tissue. Since more blood is coming in than going out, the penis swells, elongates, and stiffens. The same kind of vasocongestion happens to women during sexual arousal. Just as in males, blood rushes to the vagina and the outer area of the genitalia. The walls of the vagina become so congested that they squeeze out a fluid, a vaginal mucus that lubricates the vagina; the vagina actually sweats. The vagina also swells, becoming longer and deeper. The vaginal lubrication and the changes in the shape of the vagina are physiologically the same signs of arousal as the erect penis, they just aren't as easy to display. Blushing is another example of vasocongestion in reaction to a stimulus that is mostly beyond our control, this time to something embarrassing. If, for example, during an important meeting you spilled a glass of water over the conference table or made a stupid joke, you may suddenly feel a hot flush to your face. The brain has interpreted your embarrassment over the *faux pas* and sent blood rushing to the small capillaries close to the surface of your facial skin. Since the flow into the face is greater than the exit route of the veins, the capillaries fill up and you turn red like a beet. Vasocongestion stops when the nervous system again squeezes the arteries to a smaller size, the blood drains down to a normal level, and the redness disappears. Unfortunately for the blush, but maybe not so for the vagina or the penis, once the stimulus for embarrassment or sexual arousal has ended, the time it takes for the blood to exit and allow tissues to return to a more normal appearance is much longer than the initial rush to balloon the sensitive tissue in the first place.

During sexual arousal, while blood is rushing to the genitals, the body also responds to sexual stimuli by "tightening up"—our muscles contract and become tense, or myotonic. The same sort of body tension occurs in an animal when it is frightened by a predator and has to flee.

Sexual arousal involves other physiological responses similar to those of an animal on the run—an increased heart rate, higher blood pressure, and fast breathing. Oddly enough, these latter responses are the same responses we experience in other emotional situations such as anxiety or elatedness; as a result, mere thoughts, including nonsexual thoughts, can speed up the physiology of the body. Our bodies are telling us that something unusual is about to happen. In the case of sex, we usually don't become frightened or panicked because we notice the genital reaction as well. If the feelings of arousal that result weren't evident, if our vaginas weren't sweating and our penises weren't stiffening, the other physiological changes might send us running to the nearest tree for protection.

PERSONAL AROUSAL

The overall response to sexual stimuli, including blood rushing, muscle tension, heart rate increase, pupil dilation, nipple erection, enlargement of the breasts in women, and deep breathing, is called sexual arousal, a state that we all recognize. A person may acknowledge a tingle in the genitals or a racing heart as the first conscious clue, but each of us knows a moment of arousal when we feel it. Both men and women react just as quickly to the right set of erotic stimuli—in about ten to thirty seconds.[36] The body then becomes hypersensitive, especially in the erogenous areas such as genitals, lips, or any other place that's either physically or mentally associated with intimate pleasure, and arousal proceeds at a fast clip. It's interesting to recognize the similarities in arousal state of both the sexes. We all react to some sort of stimuli that we find erotic, and most of us respond to similar sights, touches, smells, and sounds, although there is variation in what arouses each of us. Our sexual potential is also a universal. Every person, presented with their individualized special erotic stimuli, which have probably been imprinted over a lifetime into a receptive brain, can respond very quickly in becoming sexually aroused. Humans have similar responses to sexual arousal because we are of one basic nature, one large sexual blueprint with individual variations on a theme. The stimuli might vary, but the mechanical reaction to those stimuli, with an erect penis or a lubricated vagina and a body that is poised for flight, is the same no matter who the individual is.

The Chemistry of Lust

As I turn the pages of a fashion magazine, I pause at the photograph of the young man in the Calvin Klein jeans. I see a fine male body. The visual stimulus causes my brain to fire off signals to other parts of my body, including my centers of speech, which compel me to question my eight-year-old niece, who is sitting on the couch next to me also looking at the picture. "What do you think?" I ask her. She responds, "Well, I like that chair in the picture, but turn the page." OK, no problem. My preadolescent niece hasn't noticed the lovely shoulders and the tight backside of this young man, and I surely didn't notice the chair. We're obviously on different planes here. Our difference is a simple matter of age—I am a fully mature adult female, while my niece is just on the brink of the hormonal storm that will eventually make her a sexually mature female just like me and her mother. But for now, she doesn't have the right amount of hormones for her brain to be alerted to the sexual possibilities of the guy in the ad.

THE SEXUAL ROLE OF HORMONES

Most of us know that hormones play a role in sexuality based on our traumatic experience of adolescence. The worst part of those years, as I remember, is hearing adults comment on our hormones—parents, in particular, seem to blame every unpleasant mood and every act of independence on hormones. In fact, all sorts of changes occur during the teen years. But we focus—incorrectly—on hormones as the root cause because these chemicals are involved in the physiology of what makes us female and male, and what makes us sexual adults. As the body readies itself to reproduce during the teen years, we assume hormones alone compel teenagers to spend long hours in the backseat of a car kissing each other. But strangely enough, no one is exactly sure how hormones are implicated in human sexuality.

A hormone is a chemical messenger that is free-floating in the bloodstream and changes the action of a cell. Unlike nervous signals that move quickly along direct pathways from, say, brain to muscle, hormones drift though the body, touching everything but reacting only with those tissues receptive to their powers.[37] Both women and men have all the same hormones, but we differ in when these hormones show up in the developing fetus, and how much of each hormone is produced on a daily basis.[38] When an egg is fertilized with a Y-bearing sperm, the new human will be male. During fetal development, at about seven weeks after conception, the Y chromosome initiates the production of a chemical substance which then starts the development of testicles rather than ovaries. These testicles are eventually responsible for producing the major male hormone, testosterone, which is responsible for a male-looking human. Testosterone models the growth and design of the penis and influences the appearance of characteristics that appear at puberty that are visually associated with maleness such as broad shoulders, chest hair, and beard growth.[39] If, however, the egg is fertilized with a X-bearing sperm, and there's no Y-produced substance that pushes the gonads toward testicles, ovaries appear and the fetus is female. In this case, more estrogen than testosterone is produced and those male-associated traits don't appear and the individual's gender is considered female. In this sense, the female is the basic fetal blueprint.

There are two kinds of hormones that are especially important for sexual and reproductive behavior.[40] One class, called steroids, are produced mostly by the gonads, the testes and ovaries, and are bound to plasma proteins; as a result, they are rather stable in the bloodstream.[41] The androgen steroids (of which testosterone is one) are mostly produced by the testes, and the estrogen steroids (estradiol is the most important for reproduction) come mostly from the ovaries. Testosterone is also manufactured in the adrenal glands, which lie atop the kidneys. The testosterone utilized by females mostly comes from this site. The chemical structure of steroids hasn't changed much through evolutionary time; we share the same type of these chemicals with all animals, from fish to deer.[42] This is why weight lifters are able to ingest or inject steroids made for use on horses or cows; a man's cells are as receptive to horse steroids as they would be to chemicals coming from their fellow men.

The other class of sexual hormones is called polypeptides, and they are manufactured from other glands in the body, primarily the pituitary gland. They are released only when needed to trigger specific reactions. Included among these hormones are gonadotropins: follicle-stimulating hormone (FSH), leutinizing hormone (LH), and prolactin. All gonadotropins get their names from the tissues they stimulate, the gonads. The peptide hormones are much more species specific than the steroids. The peptide gonadotropin called leutinizing hormone, for example, is important for ovulation, but the LH of a cow can't be used to stimulate the ovulatory process of a woman because our ovaries just won't recognize bovine LH.

Hormones are manufactured, secreted, utilized, and eliminated constantly throughout the body. Like nervous impulses they can also be increased by emotions or physical stress, and the production rate of certain hormones wax and wane throughout the day. Men, for example, often have an erection in the morning, and their levels of circulating male hormones, similarly, rise with the sun.

HORMONES AND SEX

Animal studies have shown a direct link between hormones and sexual behavior. Some male animals won't display any sexual activity at all if testosterone is never produced in the first place or removed experimentally from the animal's system; without testosterone, the penis becomes a sack of skin that can't respond to touch and never gets hard.[43] Many animals need some sort of hormonal activity for sex to occur. For years I've watched female monkeys mate. The group I know the best, macaques, often breed only seasonally. Breeding season usually starts in the early fall and lasts four or five months. Females in some of these species develop large swellings on their rears, signaling to males that they are fertile and would be interested in sex. Male macaques are only sexually interested in females who sport these swellings or somehow signal they desire sex in the first place. More amazing is the way the females behave as they swing those swellings through the forest. During nonbreeding season, female macaques spend most of their time with other females, usually their sisters, babies, and maternal kin, grooming each other, playing with their babies, having fights, and eating. They don't bother much with males. Suddenly, their sexual hormones change, probably triggered by the changing day length patterns of the fall, and a female macaque becomes a very different animal. As her swelling grows, she begins to notice particular males in the group. Empowered by the changes in her hormonal profile, the same female monkey who spent the last seven months avoiding large males now approaches the most dominant male she can find and presents her rear. If she doesn't conceive on the first cycle, she'll menstruate, cycle again, and continue her efforts to copulate with males she ignored a few months ago.

Hormones are the driving factor in this monkey's behavioral change toward sexuality. Her DNA has orchestrated the manufacture of those hormones to make her act that way. Because of her behavior, and the bright pink signal of her swelling, she's not only attractive to males, but interested in them herself. The change in production of estrogen produces the red hind ends and vaginal odors which signal female fertility and attract males by sight and smell. Estrogen also triggers the female motivational state to copulate.[44] Changes in hormone levels are therefore a key to initiating mating behavior in these monkeys, for both males and females.

Since humans have the same hormones as monkeys, perhaps we, too, are sexually pushed and pulled by our hormones. As in other animals, the principal hormonal system for human sex and reproduction involves the pituitary gland, the adrenal cortex, and the hormones they produce. We need a certain level of these hormones, especially testosterone, for the system to work at all. These hormones can be the key to sexual development. When boys are castrated, that is, have their testicles removed before puberty just when testosterone production is getting into full swing, these boys grow up uninterested in sex and unable to achieve an erection. But when adult males are castrated, there's no guarantee that they'll lose their sex drive.[45] Although men were often castrated in the past so that they could act as guards for harems, no self-respecting sultan could really trust his eunuchs completely. This technique has also not been completely successful in recent times. When castration was used in an attempt to decrease the libido of sex offenders it often didn't work. Nor is there necessarily a decreased sex drive in men who have had their testicles removed as the result of cancer. Subsequent hormone replacement for these males didn't necessarily restore the libido for those who had reported a decrease, either. In addition, when men lose their sex drive for whatever reason, they are rarely helped by injections of testosterone alone.

But testosterone must be implicated in some way with sex. In one bizarre study, a researcher had his wife draw his blood twice a day, first as a control and then during intercourse.[46] He discovered that testosterone levels in his blood rose during sex, but there was no way to tell if the higher level was a cause of his sexual activity or an effect. Although this study represents a sample size of only one, it does shed some light on the changes in testosterone level during sexual encounters for males.

The hormone connection is also unclear for women. Every woman eventually loses her major source of estrogen, the "female" hormone, when the ovaries shut down production as she goes through menopause. But post-menopausal women report no loss of sex drive, and in fact some say sex is even better.[47] Nor is there a loss of sexuality before menopause if a woman has her ovaries removed. Programs in which women were forced to have their ovaries surgically removed to stop them from being "promiscuous," or as a way to prevent mentally ill women from having sex and babies, were unsuccessful in reducing their sex drives. Most researchers now think that female sexuality is not an ovarian phenomenon at all, but linked to the small amount of estrogens and androgens (the more "male" hormones) produced by the adrenal glands on top of the kidneys. In a study done fifty years ago, androgen was given to 101 women who, for various reasons, were without ovaries.

Almost all the women reported that with the replacement androgen their genitals became highly sensitive, suggesting that androgens have something to do with sexual response in women.[48] The problem is that no one knows how much androgen is the "right" amount, or even if pumping hormones into those women created an unusual situation which caused abnormal sexual responses.

More to the point, the removal of testosterone, the administration of estrogen, or the introduction of androgens to either sex suffering from a reduced sex drive, or what medical authorities think is too strong a sex drive, give such variable results that no one is really sure how hormones affect our sexuality. To be sure, we need average amounts to be interested in sex or to be able to achieve an erection, but no one knows what those average amounts are. It seems that hormones play a role in our sexual interest, ability, and perhaps gratification, but that role is not straightforward. Hormones might be implicated in our sexual behavior and reproduction, but even when they're eliminated completely, or supplemented by injections, the response is unpredictable.

Our sexual arousal and activity in the end come down not to this chemical or that nerve impulse but to a combination of mind, body, and outside influences.

The Sexual Self

Everyone has a different answer when I ask them why they, or "we," have sex. "We have sex because it feels good," says one friend. Another replies, "Many women, especially young ones, do it to gain something, like affection and closeness." "For me," muses a male friend, "there are two different levels; one is intercourse, the act, and the other is sex, the whole mind and body feeling. I'm socially conditioned to have intercourse as a part of sex." "We have sex for intimacy," points out another woman friend, "and intimacy is also both physical and emotional." Although the words used are different, some of the same themes reoccur in these answers—people have sex for fun, for closeness, because our bodies have a sexual imperative. It's also clear that sex for humans isn't just an act that joins an egg with a sperm, or a drive that's mandated by chemical responses or nerve impulses. It's not just a biological response. As other researchers have pointed out, sex fulfills many other functions as well. The sex act is often how we make intimate connections with each other; sex can be a special glue that binds two people together. We murder and weep over matters sexual, or exploit others by manipulating them sexually. And in this we aren't alone. I once saw a video of two bonobos filmed at the San Diego Zoo. The male had a large leaf-encrusted branch which attracted the attention of the female. She rolled onto her back and invited him to copulate. As they mated, she casually took the branch from his hand and slid it to a safe place over her head. I've also seen male macaques attack each over access to a female in heat, and watched a dominant female monkey push a lower-ranking female away from a preferred male. In a way, it was all there—the closeness, the intimacy, and the exploitation.

Sex, like any complex pattern of behavior, has various origins and various outcomes. But most of all, each of us pays attention to, and has an opinion about, our own sexuality and the sexuality of those around us. It's both a personal act and a social phenomenon. We come to blows, I think, because sex is the only behavior that to be fully satisfying requires another person. You can find food and eat alone, you can sleep, think, and play alone. But to have a decent sexual interaction you need the enthusiastic participation of another person. And this is the problem. Each of us has a few clues about our own sexual drive and needs, but trying to figure out the sexuality of another person often seems impossible. In the next two chapters I move beyond what makes the human animal in all of us tick sexually and travel down two separate paths, one female and one male. It's not that each gender is so different, it's just that understanding the differences in terms of male and female sexuality is important when two humans with slightly different body parts meet up again in bed, or in this case, in Chapter Five.

CHAPTER THREE

The Female of the Species

In researching female sexuality, I decided to ask several friends about their first sexual feelings as illustration. To my amazement, I had a hard time deciding who might actually answer such a question. Over the years, my friends and I have discussed menstrual cramps and given each other endless amounts of advice about men, but did that give me license to confront them with the bold question "So tell me about your first orgasm?" I found myself hesitating to ask questions about a subject our culture considers highly intimate. When I finally posed the question to several women friends, they did indeed become decidedly uncomfortable, but they told me. One offered this memory: "I was very young, maybe in first grade. I remember having my hands in my crotch. You know, they fit there so naturally. And I remember my parents yelling at me to take my hands away. But it was so easy—we wore dresses, and there it was. I felt a sense of comfort by touching myself there, nothing sexual. Then later, when I was older, it went beyond comfort to something more sexual." Another told of a more harrowing experience: "When I was about twelve years old," she said, "I began to feel myself. At one point, this felt so good that I pushed something inside, a bottle or something. I don't quite remember what. I began to bleed. This terrified me and I told my mother that I had fallen on a tent pole at Girl Scouts and hurt myself and that I needed an exam. I was mostly frightened that I had somehow ruined my virginity. I was examined and told that everything was OK, I was still a 'fortress.' I was so frightened by that experience that I never touched myself again until I was thirty years old. By then I had had sex, but I was still afraid to touch myself." And then there was the woman who laughed at herself and her naïveté. "I didn't really understand much about female sexuality," she recalled, "but of course I thought I knew everything. I was eighteen, and had been having sex with my boyfriend for three months. We were in the back of his father's car, and suddenly I felt something I had never felt before. I know now that it was my first orgasm, but at the time I had no idea women experienced something like that. Suddenly I understood why women had sex." I certainly don't claim that the three stories are representative of all women in our culture; these are just a few I was brave enough to tease out of friends who were kind enough to tell me.

I knew that hearing about a variety of personal experiences would highlight something that charts and statistics don't show; not all women are alike in how, when, and especially why they engage in sex. I also had another motive—to emphasize our societal attitude toward women and sexuality. Western cultural attitudes toward women and sex are based on a Victorian mantle laid upon us by our British forebears.[1] Under that view, women were not supposed to enjoy sex, or even know much about it. According to the mores of society, only prostitutes understood the workings of sexual desire and expression; sexuality was considered the provenance of men. Although we've had a sexual revolution in this country, and women are certainly freer to express their sexuality than ever before, a double standard still envelops us.[2] Even today, it's not easy to get women in our culture to talk about sex. Joke about it? Sure. But ask serious questions about a woman's sexual history, needs, or desires, and watch her turn bright red and stutter.

In some ways, science has added to this view. It wasn't until the 1930s that physiologists clearly understood the hormonal mechanisms behind the female menstrual cycle.[3] And female orgasm is still a rather mysterious phenomenon, understood only by those who have them. In my branch of science, evolutionary biology, I see science also following what I suspect is an unconscious party line about female sexuality. This line reinforces some of those old Victorian notions from the past, but with a new twist.

Evolutionary theorists suggest that because males produce many sperm and can theoretically conceive with any number of fertile females, males should be interested in sex with anyone, anytime. Females, on the other hand, produce only a few eggs that can be fertilized and turn into offspring; in addition, females most often are the ones who bring up infants. Females then *should* be choosy about their mates; they *should* carefully select a good male and mate only enough to conceive. Afterward, they need to get on with the business of gestation and lactation.[4] We have, in essence, re-created the same double standard in evolutionary theory supplied with biological underpinnings and have applied it to all sexually reproducing animals. This theory suggests that it's natural for men to like sex with several different partners, and that it is natural that they be highly sexually motivated—it is the best way to ensure a male's genes are passed on. But according to the same theory, women should be passive and choosy, and this less active strategy is supposedly the best way for females to pass on their genes. Sound familiar?

And yet there's revolution in the air in science. We are slowly learning that this theoretical framework offered by science doesn't always fit with what females, or males for that matter, really do.[5] Recent approaches to the way we think about female reproductive biology, in particular, are changing the traditional ideas of female sexuality. New ideas on the function of how fertilization occurs, the purpose of menstruation and the function of the menstrual cycle, and the role of the female orgasm are bound to alter the traditional idea of females as passive participants in human sexuality.

The Cycle of Life

In many ways humans have disconnected sex from reproduction. But still, there's no denying that behind every sex act is the shadow of sperm and egg ready to combine and pass on genes in the form of offspring. For men, this potential for a fertilization is always at hand as sperm are continuously produced and are released as hopeful packets of genes with almost every ejaculation. For women, the connection between sex and reproduction is potentially less straightforward. A woman releases only one egg at a time, during ovulation, a process that takes place about once every thirty days. That egg is open to fertilization for only about six to seventy-two hours, at most. Throughout the rest of the menstrual cycle a woman is "infertile." This doesn't mean that she is nonsexual, however.

Other primates, such as baboons, macaques, and chimpanzees, mate only when their hormones change during estrus. The rest of the time, they're less likely to be interested in sex. Women, of course, aren't pushed to have sex by distinct hormonal cues. And yet, research over the past few years suggests that women may not be as free from their ancestral whisperings as we might think. Underlying the freedom to mate, and beneath the possibility of orgasm at every turn, is the menstrual cycle reminding every woman that the real purpose of sex is not pleasure, but reproduction.

FROM MENSTRUATION TO OVULATION

The female cycle is divided into four phases. It begins on the first day of bleeding, called menstruation. No matter how long bleeding lasts, the first day is Day 1 of any cycle, and the length of the cycle is measured from there. Around this time, the hypothalamus recognizes that a conception didn't occur during the previous cycle, and even before menstruation starts, it begins to release a hormone called gonadoliberin.[6] This substance is a peptide made by nerve cells in the hypothalamus; it acts upon the pituitary gland, the master gland found at the base of the brain right between the eyes. Gonadoliberin tells the pituitary to manufacture two peptide hormones important to ovulation: follicle-stimulating hormone (FSH) and leutinizing hormone (LH). As discussed previously, both of these hormones are gonadotropins, chemical substances that only affect the gonads, and in this case, the ovaries. During this early part of the cycle, only one gonadotropin is at work, FSH. This hormone floats through the bloodstream until it reaches the ovaries, where it attaches to target tissues—follicles on the ovaries that contain egg cells. These cells have been lying dormant ever since the woman was a fetus.

A woman is born with every egg she will ever have, but these unfinished eggs aren't mature yet and are incapable of fertilization. Instead, the eggs are arrested in their cell division, suspended in time until they're awakened many years later under the influence of FSH. This delay is caused by the special nature of the cells within the ovaries that eventually mature into eggs. All the other cells of the body, such as skin, blood, and liver cells, make new cells by duplicating and dividing once in a process called mitosis. The resulting sister cells look just like the parent cell and carry a full complement of forty-six chromosomes, or twenty-three pairs. Eggs and sperm, called sex cells, are different, however. They go through two cell divisions, a process called meiosis, to reduce the number of chromosomes in the cell from the usual forty-six to half of the normal complement, twenty-three. If they didn't, the meeting of egg and sperm at conception would be a chromosomal disaster; there'd be twice the number of chromosomes trying to entwine and the formation of a new individual would be an impossibility.

The dormant egg rests inside a follicle, surrounded by special tissue called thecal cells. These cells are stimulated to divide and multiply by FSH, and they eventually form a protective layer around the egg. No one knows why each cycle only affects certain eggs. Several follicles are, at first, stimulated by FSH, but only one usually continues to respond and produces a final mature egg. This primary follicle grows in size, fills with fluid, and can be easily distinguished from the other follicles that, for some unknown reason, have not felt the touch of FSH, at least in this cycle. The egg and surrounding tissue and fluid are now called a Graafian follicle, after the physician who first described its topography. The egg responds to all this activity by waking up from its dormant state and completing the first cell division. The thecal cells in turn start producing the major female steroid hormone estrogen.[7] Estrogen's most potent form, estradiol-17β, acts directly on the lining of the uterus. It signals endometrial cells, special cells that line the uterus, to proliferate and replace old cells that sloughed off during the last menstruation. This process readies the uterus for the implantation of a fertilized egg. Estrogen has another important role—that of ovulatory thermometer. When the follicle is large enough, and pumping out a great deal of estrogen into the bloodstream, the pituitary knows that the egg is mature enough to be released and stops producing FSH. This process takes about nine to seventeen days, starting even before menstruation. This phase of the cycle, from menstruation to midcycle, is called the follicular phase, in honor of the developing egg follicle.

When estrogen levels are high enough, the anterior pituitary releases the other female gonadotropin, leutinizing hormone (LH). The second stage of the cycle, ovulation, now begins. Right before ovulation, the egg works itself loose from the surrounding thecal cells and floats freely in follicular liquid. It has completed its first cell division and will be in the middle of the second division by the time it leaves the ovary. At this point, FSH levels are dropping and LH levels are high. The ovary acts like a cannon with a lit fuse. LH, the match, causes an enzyme reaction which literally eats away the wall of the follicle. As the wall thins out and tears, the pent-up follicular fluid and the egg along with it blast out. Some women report pain midcycle, called "mittelschmertz." It's probable that the sensation is caused by the follicle wall breaking, or the fact that fluid is now rushing through the fallopian tubes, structures about the size of a strand of spaghetti—not exactly designed to tolerate a rush of fluid. The ovary isn't directly attached to the fallopian tubes, but stringy fingers on the open ends of the tubes draw the egg down the corridor.

Meanwhile LH encourages the empty follicle to form a new endocrine structure, the corpus luteum, or yellow body. The corpus luteum now pumps out another female gonadotropin, progesterone, along with small amounts of estrogen. Together, these gonadotropins inhibit the pituitary from producing much FSH or LH, and the cycle enters its third phase, the luteal phase. If conception occurs, the egg will implant in the uterus within seven days or so. If this happens, the corpus luteum has the job of producing large amounts of progesterone, which maintains the uterine lining until the fetus's placenta takes over this task at about three months' gestation. But if conception doesn't take place, the corpus luteum begins to disintegrate. As a consequence, levels of estrogen and progesterone also decline. When progesterone levels become extremely low, fluid in the endometrial tissue is sucked up by the blood and causes the uterine lining to buckle and shrink. The arteries that once connected the lining to the uterine wall close off and the top layer falls away. A shower of blood, endometrial tissue, mucus, and vaginal cells drops through the vagina. This is the fourth stage of the cycle, menstruation.

The Curse

We think of menstruation as the end of the cycle, something sad and disappointing, something to be ashamed of, perhaps. But one of the recent revolutions in female sexual biology offers an alternative interpretation of menstruation, viewing the end of the cycle as something positive rather than something negative.

Although many women don't understand the process of conception, or how exactly the cycle moves from menstruation to ovulation, there's no denying the flow of blood from the vagina at menstruation. There is no such thing as a "normal" menstrual cycle length. A thirty-year study of 250,000 menstrual cycles of 2,700 women showed that only one woman had "normal" twenty-eight-day cycles during her menstruating lifetime.[8] In fact, human menstrual cycles are characterized not by regularity, but by their variability. Post-adolescent cycles are irregular and often without ovulation. Women who nurse have their cycles interrupted, as well. As one approaches menopause, the cycles take on another pattern. Sickness, weight loss, and extreme stress can all influence any particular cycle, and certainly every woman has had any number of these events in a lifetime. As a result, no woman can chart an even path through her menstrual life history. But the impact of menstruation as a biological and social fact is imprinted on our species.

THE CULTURE OF BLOOD

In my living room, I have the kit of an adolescent !Kung San girl from Botswana. It includes a small leather apron to cover her genitals, a pouch of powder, and a special stick which she must use for eating. Although this kit might be considered a piece of ethnic oddity, it's also not much different from the gear given to American girls when they first menstruate—a sanitary belt, pads, some aspirin, and a calendar to keep track of their periods for the rest of their fertile lives. These items, too, enroll a young girl into the secret society of womanhood. I remember being told, from the time I first menstruated, not to go swimming when I had my period. I thought this was because the bleeding would cause the same sort of cramp I had been warned would happen if I dared venture into the water too soon after eating. Many years passed before I realized that the swimming taboo referred to the embarrassment of a huge wet sanitary napkin poking out of my bathing suit; with tampons I was free to swim, ride, and hike, just as the advertisements promised. Our culture, then, has created a special aura around a woman's period.

Most cultures define womanhood from the moment of first menstruation, called menarche, and many mark this occasion with some ritual. In some cultures, girls are sequestered away from the community, allowed to eat only certain foods, and bathed in certain ways. Some taboos continue past the first menses. It's common for women in some cultures to be forbidden to interact with men's hunting tools, or to stay away from water sources. Menstrual blood is often thought evil and dangerous, and by implication, powerful.[9] In our culture, some women see this rhythm as the defining feature of femaleness, and encourage women to see power, rather than weakness, in the process of menstruation.[10] Other societies still call it a "curse." In a 1981 World Health Organization survey of British women, 54 percent said that intercourse should be avoided during menstruation.[11] But it seems that our universal repulsion at the bleeding of menstruation might be wrong. We should be celebrating menstruation—not only, as some feminists suggest, as a sign of our femaleness, but also because it might be Nature's way of guarding us against infection as we go about our sexual business.

THE END OF "THE CURSE"

In the summer of 1993, biologist Margie Profet proposed a theory that might give menstruation some other moniker than "curse." Profet feels instead that menstruation is not a curse, an evil, nor even a way for the body to slough off dead tissue. She suggests instead that menstruation evolved as an adaptive response designed to rid the female body of pathogens that ride into the vagina and uterus on the backs of sperm.[12] These bacteria are implicated in vaginal and uterine infections, and infertility, and present a major sexual cost to sexually active women. Menstruation, which drags out these potentially infectious bacteria, would then be an adaptive response that evolved to help women fight off the dangers of sexual intercourse. Biological evidence to support this more adaptive hypothesis are the specialized spiral capillaries in the uterine lining which aid in cutting off blood supply; the nonclotting character of menstrual blood, which makes it flow easily; the fact that menstruation occurs regularly in species where females copulate frequently, thereby exposing their internal organs to pathogens; the fact that menstruation is not limited only to human females but seen in other animals as well. She also questions the old view of menstruation as a "sloughing off." Why would natural selection, she asks, promote such copious blood loss simply to discharge a disposable lining? Instead, Profet sees menstruation as an anti-pathogen response, an antibiotic adaptation that evolved to help females charter their sexual lives through a maze of males with bacterially infected sperm.

There are problems with this concept, of course. Most important, repeated menstruation is a product of our contraceptive age. For most of our history, women reached menarche much later than we do now, around fifteen or sixteen. They also experienced at least three years of adolescent sterility, followed by a cycle or two, and then a pregnancy. Lactation went on for years during which there was no menstruation, but presumably sex, and once ovulation began again, pregnancy soon followed. As a result, women used to have many fewer periods than we do today, and thus Profet's idea of menstruating as a way to rid the body of pathogens couldn't have been much of an advantage.[13]

Beyond that important criticism, this idea is still remarkable for two reasons. First, it has the potential to overturn the way we think about a female universal. No longer can we claim menstruation is an awful event that occurs only because the uterus needs to be renewed for a possible future conception. We have to throw out our old notions of menstruation and its mechanisms and reevaluate it as a possible adaptive response. Second, and more important, this study adds to what I see as the ongoing revolution in female sexuality. This work says, in essence, that female reproductive biology is not a passive event. Menstruation, rather than a negative effect, is an active mechanism to protect a sexually active female body against disease. This is not a matter of loss and passivity here, but a finely tuned adaptation.

In any case, there's no denying the important place of the female menstrual cycle in the reproductive and, by implication, sexual biology of women. The next evolutionary question is how the cycle influences how women go about their sexual business.

Cyclic Control

A person can temporarily control their breathing. They can decide whether or not to urinate or sleep. Many of our body functions have both an automatic response system as well as a gray area in which we are able to exercise some control. We can't really control our heartbeats, but on the other hand, we can make our hearts race by thinking about someone we love and we can make our heartbeats slow down by sitting peacefully while thinking about a tranquil day at the beach. In the same fashion, no woman controls her menstrual cycle length, or when she ovulates, or how long menstruation lasts. But there is a combination of influences that has the potential to change the cycle.

SISTERHOOD IS POWERFUL

Almost every woman I know ascribes to the idea that one woman's cycle can be driven or influenced by another's. And science has evidence suggesting that they're right. I have one of these stories too, but it has a different kind of twist.[14] Several years ago, I was attending a conference in upstate New York. Two of us had rented a hotel room at the conference headquarters, but since all our friends were impoverished academics, we decided to sneak a few more colleagues into the room and share the cost. As a result, our room eventually slept seven people—on the floor, in beds, in chairs. All but one of our roommates were women; the one male was a married friend, a man very much at ease in a room full of women. That first night, after much discussion about the lectures we'd heard that day, and a little friendly gossip on the side, we turned out the lights and settled down to sleep. Just as we were drifting off, the lone male voice in our group made the comment: "Gee, I feel my cycle synchronizing." He couldn't have had a better audience—a room full of women scientists knowledgeable about female reproductive biology. We laughed as only people who share an inside joke can.

All of us were well versed in the work of Martha McClintock, a professor of psychology at Harvard, who in 1971 published a study on the cycles of 135 women living in dormitories. After a few months, women living close to each other or spending more time together started to have matching cycles; roommates and close friends eventually menstruated about the same time and had cycles of about the same length.[15] Ten years later, this work was confirmed more specifically for roommates, close friends, and prison cell mates, that is, women who spend a lot of time together.[16] Since best friends synchronize more than roommates, the effect seems to be the result of intimate attachment more than mere proximity.[17] Cycles become synchronous in the sense that menstruation appears about the same time, and by implication this means that they're probably ovulating about the same time too.

How does this happen? The general consensus among the scientific community is that it's chemistry. In one study, for example, a woman with regular cycles was asked to wear a cotton pad in her armpits all day.[18] The pad was then cut into quarters and the pieces were frozen in dry ice for a few hours. Five other women consented to have this cold pad wiped under their noses three times a week for several months. At the start of the experiment, the subjects' menstruation cycles were on average 9.3 days away from that of the donor, but after a few months, four of the subjects synchronized exactly with the donor, and the fifth's cycle was close. Since none of the subjects reported smelling anything on the pad, scientists concluded that nonodorous chemicals, or pheromones, were at work. These pheromones, it has since been discovered, aren't restricted to the armpits. The vaginal wall produces aliphatic acids, too, that change with cycle state and are released into the air. Under the influence of estrogen, they're potent during the late follicular stage, close to ovulation, and decline soon after ovulation.[19] The possible role of free-floating chemicals is confirmed by the fact that women synchronize more when they use sanitary pads rather than tampons during menstruation, thereby letting their pheromones fly free.[20] The point is, women in rather intimate contact, in a position to be swamped by chemical fumes for several months, synchronize their cycles like the kicking legs of the Rockettes.

The evolutionary significance of synchronized cycles is unclear. One possibility is that long ago when ancestral women cycled the same, men would know exactly what was happening and could track them more easily.[21] This is probably not a very reasonable hypothesis because women don't evolve to accommodate men. Another possibility is that when women cycle together, they will also conceive together and give birth together. A test using vaginal smears of ovulating women and seeing if those smears might influence ovulation in other women would be useful. If women did ovulate and conceive together, such synchrony might have promoted shared infant care in our past, or perhaps their common fertility increased male-male competition.[22] But since we know little about the communal living conditions of our ancestors, it's still a mystery why some women are able to have such a driving effect on their close associates.

SWEATY T-SHIRTS HAVE A USE

There's another side to the synchronization story. The mere presence of a man can also change a woman's cycle. In one study, when women spent two or more nights per week with a man, regardless of their sexual interaction, they ovulated with more regularity than those who never, or more sporadically, slept next to the opposite sex.[23] To test the possibility that the scent of one's companion might influence the cycle, researchers again turned to the chemicals emitted by the body through the armpits. They recruited men, this time, to wear cotton pads in their armpits.[24] The researchers described their subjects as men with "large numbers of lipophilic diphtheroids in [their] axillary region" and a full range of armpit odorants full of volatile fatty acids and steroids; these were sweaty men. Liquid from the pad was distilled and rendered nonodorous. This time, fifteen women with irregular cycles were the targets; male sweat extract was wiped on their upper lips three times a week. After a few weeks, the cycles of these women became more regular and less variable; they seemed to stabilize in a menstrual sense. The obvious conclusion is that the close, intimate presence of a man, or at least his armpit smell, helps keep a woman's menstrual cycle regular. The evolutionary advantage of such an influence might be that maleness will regulate the cycle and possibly make ovulation and conception easier. Certainly, such entrainment suggests a cooperative reproductive effort within couples, rather than a competitive interaction.

MORE THAN SWEAT GLANDS

The act of sex, it appears, is also a significant modulator of female cycles. Several studies have compared the sexual behavior of women and the length and regularity of their cycles.[25] It turns out that women who have sex one or more times a week tend to have average-length cycles, cycles we would label as regular. Sexually active women also seem to ovulate more often than women who have sex only occasionally or not at all.[26] The implication is that regular sex with someone will result in more regular cycles. Does masturbation serve as well to keep the cycle regular? Apparently not. The same research shows that masturbation has little or no effect on menstrual cycles. Studies of 1,066 women over a two-year period in five countries shows that when intercourse occurs with regularity over both the follicular and luteal phases, the cycle swings into a rhythm.[27] Oddly enough, the effect of regular sex can't be charted in smoother hormone levels of FSH, LH, or estrogen. In fact, no one is really sure how intercourse affects the cycle. Scientists only know that the effect is cumulative over the cycle, and women do best, in the ovulatory sense, when they have regular weekly sex. This doesn't mean that ovulation is induced by copulation, as it is in cats; but it does suggest that rhythms can be improved, changed, and altered slightly by outside forces.[28]

The bottom line is a bit fuzzy here; these studies don't really tell us if such chemistry helps in the ultimate evolutionary purpose of passing on genes. They imply that cycles are more regular in the intimate presence of men, other women, and with an active sex life. But does this mean women under these influences conceive better than less regular women, or produce more infants? No one knows. This work does suggest that our social environment—living with other women and with pair-bonded male partners—has an influence on female reproductive physiology. No woman is an island, at least in the sense of menstruation cycles. These data also add a new wrinkle to the theory that ancestral human females used sexuality to keep a male close to home. Perhaps pre-hominid women didn't have sex with males just for the paternal care that such seduction might bring, as some have suggested (see Chapter One). Another possibility is that ancestral females had regular sex with males because it kept them reproductively on key. With regular cycles, these females would ovulate better, conceive more quickly, and stay menstrually regular at times when they were not pregnant.

What's also important is how those regular ovulating cycles intersect with the act of sex. From an evolutionary point of view, a regular cycle is only a positive benefit if it results in conception more easily. And for that, we need to see if women are driven to have sex because of the rhythm of their cycles.

The Cycle of Sex

We assume that our species is less driven to reproduce by hormonal influences than are other animals. For example, we assume that women, unlike monkeys in estrus, aren't driven by their hormones to have sex. Although female hormones drive the menstrual cycle which matures an egg, we believe these changes shouldn't affect a woman's sexual desire. A human female's sexual interest, therefore, should be the same over the cycle length. More important to the study of human evolution is the assumption that this "flat line" reflecting human female sexuality has to be explained, because it's so different from the sexual fluctuations of our primate cousins who adhere to sexual interest only when it's necessary—around ovulation. Two decades of research on female sexual behavior, however, indicate that our thinking on this issue is all wrong; human females might be more like their monkey and ape cousins with regard to sexuality than many of us would like to believe. Many women report, as their primate sisters would if they could talk, a surge of sexual desire right around the time when they ovulate.

MY HORMONES MADE ME DO IT

The first study to look at the hormones of the ovulatory cycle and sexuality was conducted in the 1930s by two psychiatrists.[29] They studied a group of clinically diagnosed neurotic women by correlating their moods, judged by the psychiatrist from therapy sessions, with the amount of estrogen in smears taken regularly from the women's vaginas, and drew conclusions about their subjects' sexual attitudes. The psychiatrists concluded that female sexual behavior is correlated with estrogen, but their work is obviously flawed, because it evaluates emotionally unhealthy women. In addition, vaginal smears aren't the best indicators of estrogen in the blood. Correlating hormone levels and sexual behavior has been refined in recent years with radio immunoassays that use tiny

amounts of blood to get accurate hormone levels from the subjects involved. But even with these more sophisticated methods, no one can
show that estrogen has much to do with mood changes and practically
nothing to do with sexual interest.[30]

This is no surprise. As I mentioned earlier in Chapter Two, estrogen is
a reproductive hormone that probably has only a slight influence on
sexual motivation. Better studies measure androgens and their role in
female sexuality. In one work by Persky, women with higher levels of
testosterone in the blood during ovulation reported more frequent sex.[31]
Interestingly enough, Persky also found that women who have what he
classified as higher levels of testosterone were also less depressed, found
sex more satisfying, and said they were able to form relationships more
easily than women with lower levels. This suggested to Persky that
women with higher levels of testosterone would be more motivated to
form pair-bonds and their sex lives within that bond would be better.

Although this is an interesting scenario with important evolutionary
implications, other researchers have contradicted Persky's findings.[32]
More important to an understanding of cycles, hormones, and sexuality,
testosterone isn't one of the reproductive hormones that waxes and
wanes relative to the female cycle. It surely has an effect on motivating a
woman to think about sex and approach a partner, but this probably has
little to do with having sex at the "right" time for conception.

And so we are left with a hormonal conundrum. Estrogen, which is
closely linked with ovulation and the cycle, doesn't influence sexuality,
but testosterone, which isn't cyclic, does. Perhaps the point is not to
wallow in the hormonal link between estrogen and estrus that drives the
nonhuman primates to mate, but look at what women actually do as
sexual beings, hormones or no hormones.

SEX AND THE OVULATING FEMALE

Even if we don't know the exact biological influences underlying female
sexual motivation, scientists suspect that left to her own devices, a
woman will express her sexuality only when it suits her purposes. From
an evolutionary perspective, it makes sense for a woman to control her

sexuality. Women can be sexually attractive to men and be interested in them during any time of the cycle, but women also have the option to have more, or less, sex at different stages. Given that ability, it would make sense in evolutionary terms for women to be highly sexed at ovulation. What better way to pass on genes than to have an innate drive for sex at ovulation? And so scientists asked the simple question: Do women have sex more often when they ovulate?

The first reference to the distribution of sex over the female cycle comes from a self-help book by Marie Stopes published in 1918. Dr. Marie Stopes was a lecturer in paleontology at Manchester University, held a doctorate in science, and published several books on plant life. Apparently, she was also the Dr. Ruth of her time. She published books on motherhood, birth control, and conception. In her volume *Married Love: A New Contribution to the Solution of Sex Difficulties,* she addressed the female sex drive, or as she called it, the "fundamental pulse."[33] She tracked this pulse in married women who were, for business or professional reasons, separated from their husbands by asking them to keep track of when they felt the "urge" for their mates. Based on information from these women, the pattern of sexual desire is clear; the women with absent husbands wanted sex most often at midcycle, and then again right before menstruation. There is, Stopes maintained, a "spontaneous sex-tide" in all women.

What is remarkable about this book, and Dr. Stopes's information, is that the link between ovulation midcycle and menstruation was as yet unknown. And yet here is clear evidence that when asked, these women admitted to desiring sex more often at midcycle than at any other time. Other studies during this time by psychiatrists also confirmed a peak in sexually oriented dreams and feelings during midcycle.[34] A more "scientific" study was conducted in 1937 to check the physical and emotional patterns of women over the cycle.[35] The researchers in this study found no pattern in the distribution of sexual encounters during 780 cycles of 167 married and unmarried women. But they did see a rise starting about the eighth day of the cycle in what they defined as female sexual desire. The study implies that while women might want sex more during this time, that doesn't mean they actually have more sex.

From the 1960s on, there's been a flurry of research on when exactly

WHAT'S LOVE GOT TO DO WITH IT?

women have sex. I think it's reasonable to suggest that the feminist revolution in this country was a major factor in shining a spotlight on female sexuality research. In several studies, women were asked to keep track of their intercourse and orgasms on daily charts. The statisticians played with this information to determine if there was any sort of sexual peak during the time of probable ovulation. These studies are fraught with all sorts of problems. Do you pinpoint ovulation by counting forward or backward from menstruation? Since there's no such thing as a regular cycle, how can you compare the experiences of women as a group, or any single woman over time? Can you be sure a woman really ovulates? Will people accurately report about their sex lives? The results of these studies are predictably mixed.

Some work clearly shows a midcycle peak in sexual activity and indicators of heightened sexuality, such as orgasm frequency, female initiation of sexual intercourse, and masturbation.[36] Other studies are not quite that clear, but do show at least a rise in frequency at midcycle.[37] It looks like the mechanism for this midcycle peak might be testosterone, which sometimes shows a slight midcycle peak as well.[38] Since female sexuality, or male sexuality for that matter, is always compromised by what one's partner might need or desire, scientists, in examining the effects of sexual activity and cycles, have factored out male responses and looked primarily at the desires of women alone. The suggestion is that if women ruled their sexual lives and didn't have to compromise with men, they would most often have sex when they were ovulating. One imaginative study dismissed the influence of males by looking only at lesbian couples and their pattern of sexual interaction.[39] These women, who were not compromised or dominated by male sexual pressure, clearly had a midcycle peak in initiation, total sexual encounters, and the frequency of orgasm. They also didn't show the typical premenstrual rise of sexual frequency shown in their heterosexual sisters, nor did sex decline during menstruation, as it does in other women. But a midcycle peak in female desire appears in the studies even when men are involved with the women subjects. And when there is a pattern, women most often report a sharp rise in sexual desire, initiate more sex, reject partners less, and masturbate more frequently during ovulation, or close to it. This heightened sexuality sometimes reoccurs right before menstruation.[40]

But there are also studies that show inconsistent patterns of sexual frequency, or no patterns at all.[41] Some report a pattern of desire that is linked only with menstruation rather than with ovulation, and this pattern presumably has no evolutionary effect.[42] This is especially true for studies that have looked at female arousability. The assumption is that if women are more attuned to sexuality at midcycle because their hormones are pushing them to have sex, they would also be more easily aroused. In several experiments, scientists attached small thermometers to the labia of female subjects and inserted a device called a photoplethysmograph into their vaginas to gain physiological measures of sexual arousal.[43] Presumably, the labia would heat up as the women became excited and the photoplethysmograph would chart vasocongestion of the vagina; both are quantifiable measures of a woman's arousal state. Women in these studies were wired up and then exposed to erotic audio or visual tapes and their arousal states were measured. It seems that no matter where in the cycle a woman is, women get aroused at about the same level.[44]

This tells us more about female receptivity, to use the terms better reserved for species with clear estrus, than about a woman's proceptivity, or her motivation to seek out sex. Despite the fact that a woman can be stimulated at any stage in her cycle to enjoy a sexual encounter, it doesn't mean much in an evolutionary sense. The point is, when a woman is in full control of her sexual interest and arousal, she will *usually* be more interested in sex at midcycle. Without the use of birth control, without the complications of male demands, a woman will be compelled by her basic nature to mate during the time when conception is most likely. This doesn't mean that she won't have sex during nonfertile times of her cycle, and this is where human females are different from other animals. It just means that women, like other animals, are biologically and hormonally motivated to have sex most frequently at ovulation to ensure conception.

What's more difficult to explain is the way society uses this connection between hormones and the menstrual cycle to govern women's lives. We live in a society that has kept women out of powerful positions and out of dangerous situations such as combat with the excuse that women are "more emotional" and more "swayed by their hormones." The reasoning goes something like this: The menstrual cycle is a process of great hormonal changes. Hormones influence mood changes; as a result, women are subject to uncontrollable moods, and by implication, those mood swings are negative. What science tells us is that yes, some women do change their behavior and subconscious thoughts relative to the menstrual or ovulatory cycle. But in the case of ovulation at midcycle, this has less to do with moodiness and more to do with a desire for sex.

The Peak Sensation

Try to describe the taste of a chocolate bar. It's sweet, it's smooth, it's rich, and well, it just tastes good. It turns out that it's almost impossible to describe a taste in a way that accurately explains to the reader why someone might crave this substance. The same is true in trying to describe orgasm. Clinically, orgasm is described as an intense sensation marked by physiological changes in the body such as muscle tension and a feeling of disconnection, and then a release from that intensity.[45] Of course, this description is not particularly illuminating. It's like saying chocolate tastes good. And no wonder there's no perfect description; how can a person describe something when they aren't fully conscious of the feeling when it's happening? Orgasm is a moment of great vulnerability for individuals. For the few seconds of its duration, your mind and body are completely occupied, completely disconnected from the outside world. Perhaps it is the moment we are most connected to our primitive selves. People cry out, moan, and make funny faces, and in that we are more animal than we might like to acknowledge.[46] For men, orgasm includes the release of seminal fluid and sperm—and a chance to pass on genes if he's having sex with an ovulating woman. For women, orgasm itself is more disconnected from reproduction. And yet the female orgasm, like the male orgasmic release, may have also evolved over time to ensure that women participate in sex in the first place.

THE FEMALE TRIGGER

There is no record of what pre-human cultures thought of female orgasm. The description of female orgasm in modern Western culture has been affected most by Sigmund Freud's writings from the turn of the century.[47] Freud believed, based on his sexually dysfunctional female patients, that female babies center their sexual experience on the clitoris. As girls grow into adults, he reasoned, this pleasure must transfer from the clitoris to the vagina; according to Freud, a sexually healthy woman should have only "vaginal" orgasms.[48] For the first half of this century, women in Western culture were told that if they didn't experience what was known as a "vaginal" orgasm, there was something decidedly infantile in their sexuality. Many women reacted by searching in vain for the illusive deep orgasm they had read so much about. It wasn't until the 1950s and 1960s that scientists broke away from the Freudian concept of female orgasm and began to ask women themselves about their bodies. Although women, and men for that matter, experience all sorts of intensities in orgasmic pleasure, scientists now know there is no difference between a clitoral and a vaginal orgasm. There's only one general kind of orgasm and it starts with that bundle of nerves at the top of the female genitalia.

The clitoris, unlike any other organ or tissue in the body, exists only to receive and give off sexual pleasure.[49] The clitoris is a boneless mass composed of two stems of tissue, the corpora cavernosa, which respond to sexual excitement by filling with blood and expanding. In an anatomical sense, the penis is much like the clitoris in its ability to enlarge as a result of becoming engorged with blood. The tip of the clitoris, the glans, is especially sensitive, much like the glans or head of the penis. It's full of sensitive nerves including those that respond to pressure, temperature, and touch. These nerves bundle together with a large dorsal nerve, and eventually join with the pudendal nerve running to the spinal cord at the lower back. A section of the labia minora, the inner, smaller lips of the external genitalia, folds over the clitoris as a hood. The clitoris is set above the vaginal opening, in front of the urethra; there's no direct connection between the clitoris and the vagina. Just as in all parts of our bodies, be it breast size, penis size, or eye shape, there's no such thing as an "average"-sized or -shaped clitoris. Some are larger or smaller, and some are positioned higher or lower depending on the origin of the ligaments which hold the clitoris to the front of the pelvis. The body of the clitoris can be short or long, and the glans, too, can be small or large. If this description of variability sounds faintly familiar, it is because the penis is an anatomical homologue to the clitoris. But the similarity stops there. In men, the penis reacts first to sexual stimulation, but in women, the vagina, not the clitoris, reacts first by producing lubrication. When the clitoris does respond it doesn't really become erect, as the penis does. While the clitoris therefore is anatomically a homologue to the penis, it is not necessarily similar in its function and response.[50]

THE FEMALE RESPONSE

Masters and Johnson describe female sexual response by dividing the sexual progression into four phases—excitement, plateau, orgasm, and resolution.[51] But don't be fooled by this systematic classification. Women need not follow a straight line progression when engaged in sexual behavior, and they can move from one stage to another again and again, or skip a stage. The point is that female sexuality is distinguished more by its flexibility than anything else. It starts with some kind of sexual stimulation—an erotic thought, a memory, some physical initiation, or sensory input involving sight, touch, or smell. Within seconds, the vagina begins to lubricate, and if the process isn't interrupted, blood rushes to the genitals and swells the labia minora and makes them deep red.[52] The position of the uterus and the shape of the vagina also change as the whole pelvic girdle alters with mounting excitement. The vagina opens up and the inner two third stretches. The uterus is pulled higher into the body, lifting the cervix, the opening to the vagina, away from the end of the vagina. The clitoris also begins to respond by increasing in width (but not in length as a penis would). The end of the clitoris, the glans, swells as well. The outer lips of the vaginal opening, the labia majora, also become engorged and flatten somewhat against the pelvic floor. Other parts of the body, too, join in the sexual excitement; the nipples become erect, breast size gets larger, the body tenses, and heart rate increases. Some women become flushed all over as their body temperature rises. If the process continues, the excited woman enters into what Masters and Johnson have labeled the "plateau phase." This phase is the staging platform for orgasm. The body at this point continues on its path of muscle tension, heart rate increase, high blood pressure, and deep breathing.

Orgasm is expressed all over the genital area; in preparation, the clitoris retreats under its fold of skin, the clitoral hood, and the outer third of the vagina swells tight. The pressure of a penis going in and out of the tightly swollen vaginal opening then acts indirectly on the clitoris by pulling down the sides of the labia. This stage is called the "orgasmic" platform, and sex researchers believe women who say they don't have orgasm simply never reach this level of excitement. At this point, the genital area is so swollen that applying any pressure on the clitoris is a simple matter of push and pull on any place in the area. The body is now ready for the "orgasmic phase."

Physiologically, orgasm involves strong muscular contractions in the outer third of the vagina, contractions of the uterus, and echoes of those contractions around the rectal sphincter.[53] Interestingly enough, women subjectively report they are having an orgasm before the contractions really start. The contractions can be regular or irregular, show up in any set number, and last for varying lengths of time. Although the clitoris, the labia, and the breasts don't change much physiologically during orgasm, the rest of the body is working overtime. The heart rate might double and blood pressure is commonly up a third. Some women also emit fluid during orgasm that has been mislabeled as "female ejaculation."[54] Although there hasn't yet been a good anatomical study of the phenomenon, the fluid appears at orgasm in an estimated 40 percent of all women. It supposedly comes from multiple glands, called Skene's ducts, near the urethral opening. These ducts vary among women in their size and secretory volume.[55] They're remnants of the embryonic tissue that in male fetuses develops into the prostate gland. In women, the fluid from these glands empties into the vagina at orgasm and is often mistaken for a flood of vaginal fluid, or as an embarrassing urinary emission. Female ejaculation is also associated with an area on the anterior vaginal wall called the G-spot, which, in a very few women, is highly sensitive because of its close anatomical association with the urethra and when directly stimulated can bring on orgasms quickly.[56] Women describe the moment of orgasm as a feeling of being suspended from their bodies; others find themselves entirely focused on their pelvic areas, and all other conscious thoughts disappear.[57] In all cases, the feeling of orgasm is one of extreme pleasure.

Women don't always move back to a resting state, called "resolution phase," directly after orgasm. Sometimes they subside back into the plateau stage, and can be called to orgasm again. This is a clear difference from the male orgasm; for men, the time after orgasm is one of falling back directly to resolution phase, where they must stay for several minutes, even hours, before they can be sexually aroused again. Although a woman can achieve orgasm immediately again, more often she slides into the resolution phase and reverses all the physiological signs of sexual arousal. Blood leaves the vagina, the vaginal lips, and the clitoris; the clitoris pops out from the hood and is exposed once more; heart rate and breathing return to normal.

The idea of a clitoral versus vaginal orgasm for women is out of fashion these days, but some scientists still like to classify and catalog female sexual response. For example, one researcher differentiates between orgasms that stimulate the orgasmic platform which can be achieved by masturbation, and orgasms that involve the movement of the uterus and repeated poking of the cervix, which of course requires the insertion of something.[58] These researchers contend that since women describe these orgasms differently, there must be two distinct kinds of orgasm. This is like suggesting that since men describe masturbation and intercourse differently, they have two distinct types of orgasms.

While women can have multiple or serial orgasms, this doesn't mean that all women do, or even that some women do regularly. When women do report multiple orgasms, they aren't described in the same way as single orgasms. Sometimes they are described as a rapid series of intense waves right at the level of orgasm, or a distinct return to the plateau level followed by a rise again to full orgasm.[59] There's some evidence to suggest that women who experience multiple orgasm regularly are also more sexually experienced and involved than other women. But no one knows if these women engage in more sex because they are multiorgasmic or vice versa.[60] In any case, the basic stages of excitement, plateau, orgasm, and resolution are similar in men and women. In addition, each person has all sorts of orgasms, varying considerably from mildly pleasurable to intensely satisfying.[61]

WHOSE ORGASM IS IT ANYWAY?

We know that men have to have orgasms to expel sperm; without the orgasm a man has no chance at passing on genes. But the female situation is quite different. No woman needs orgasms to expel an egg, and certainly orgasm isn't necessary for conception. Some biologists think that the search for an evolutionary explanation for the female orgasm is a completely useless exercise in the first place.[62]

When a fetus is first developing, it is sexually undifferentiated for a few weeks until, under the influence of testosterone, the chromosomal male fetus begins to differentiate itself and develop testes and a penis.[63] Some scientists have suggested that the female orgasm is simply a byproduct of this shared fetal history.[64] The same reasoning can be applied to explain the presence of male nipples. Women need nipples for nursing, and since we all come from the same general blueprint, men have them too.[65] But male nipples are basically dysfunctional dots on the chest that never give off any milk, while the female orgasm is a powerhouse of pleasure.[66] And female orgasm has an evolutionary history all its own that can't simply be brushed off as a byproduct of maleness.[67]

Some primatologists think that wondering about the origins of human female orgasm is unnecessary in the first place for a different reason. Experimental work on macaque monkeys has demonstrated that human females aren't the only ones who orgasm—certainly macaque females do as well. And so it might not be a special feature of humans requiring a special explanation.[68] Other primate females have clitorises, sometimes huge ones that they rub on each other, on males, or branches to gain pleasure.[69] The clitoris appears in some reptiles, birds, and often in other mammals.[70] The female clitoris, the pleasure it brings, and the expression of orgasm must be part of our animal heritage, specially evolved for some biological reason in the female line.[71] If female orgasm is a major feature of human female sexuality, why do females have them?

KEEPING HER DOWN ON THE FARM

If female orgasm has adaptive value, there must be some evolutionary explanation for the selection of female pleasure. It might be that orgasm simply functions to make sure women stay interested in sex. Any intermittent reward can be used to make someone return again and again on the off chance of obtaining a pleasurable goal, and those who do so will have more sex more often and be more likely to conceive.[72] Occasional orgasm with a male might also strengthen the evolution of the human mating pattern, the so-called pair-bond. One biologist has even suggested that a repeated reward of orgasm with a particular male will help a female choose her best long-term partner.[73] Under this scheme, if a women has orgasms it will bond her with a permanent mate, and provide what one author calls a "domestic bliss function."[74] If so, it isn't working very well at the moment in many societies. Today, women in Western society report they have fewer total orgasms than men, and that at least 50 percent of all orgasms occur without a male partner.[75] And since women are aroused just as quickly as men when they have the right stimulus and follow through, the lower rate of copulatory orgasm in today's women might just be a product of less accomplished partners and bad sexual technique.[76] In other words, the lack of orgasm has just the opposite effect on the pair-bond as it sends women off to seek satisfaction in extra-pair copulations.[77] Or perhaps the 50 percent rule is simply an artifact of our repressive culture. In societies where both women and men are expected to orgasm, the rate of female orgasm is high, but when female orgasm is ignored as a "natural" sexual response, and men pay no attention to arousing their partners, the rate is reportedly low.[78]

If the pair-bond doesn't seem like a compelling explanation for the evolution of female orgasm, what other evolutionary advantages could select for female orgasm?

THE HOOVER THEORY

Anyone who's had heterosexual sex knows that one common result of intercourse is a wet spot later on the bed. This pool of wetness is the result of a mixture of seminal fluid, sperm, and female excretions leaking out of the vagina sometime after ejaculation. From the male's evolutionary point of view, this "flowback" is a disappointment. It means many of the male's sperm have been ejected and have no chance of fertilizing the egg. One possible theory to explain female orgasm is that a woman who has had an orgasm might be more likely to stay horizontal longer, retaining the maximum amount of sperm in her vagina after intercourse.[79] Also, once a woman orgasms, her cervix returns to its previous position, dipping directly into a pool of semen left as a reservoir in the upper third of the vagina.[80] But it doesn't make evolutionary sense that women will evolve the capacity to orgasm and stay flat because their orgasms are advantageous to men.

It's more likely that the female orgasm evolved to help females sort through the tangle of male reproductive tactics. When orgasm occurs, the uterus contracts in a rhythm and acts like a vacuum, sucking up sperm. This has been confirmed by a study in which a device was placed into the vagina of a woman to measure barometric pressure.[81] Vaginal pressure rises during female orgasm but falls when a man pushes in his penis and orgasms. It rises once again once a woman relaxes after orgasm when the pressure decreases; sperm could theoretically be sucked more deeply into the vagina and possibly uterus at this point.[82] But to consider seriously the idea that female orgasm may have evolved to help control sperm intake, we need to show exactly what happens to sperm during the process of orgasm.

Two British biologists known for their work on human sperm, Robin Baker and Mark Bellis, recently conducted a study of the female orgasm as a possible tactic to manipulate sperm.[83] They hypothesized that a woman can influence how much sperm is allowed into her reproductive tract in several ways. First of all, she can refuse to mate. More subtly, she can also stand up after sex and let most of the sperm inside her vagina leak out. She might also strategically orgasm and thereby manipulate the flow of sperm in and out of her reproductive organs. Baker and Bellis gathered data from women in various ways. Several female subjects agreed to collect flowback, the fluid that leaks out of the vagina up to thirty minutes after sexual intercourse. This information was supplemented with sperm counts of ejaculations gathered in condoms, and by lengthy questionnaires in which they asked their subjects all about their sex lives. Baker and Bellis discovered that women often masturbate to orgasm in between sex with their partners, and that at least 35 percent of copulations are nonorgasmic for women. In addition, human females retain only about 65 percent of an ejaculation most of the time. They also have the potential to get rid of the entire ejaculate, sperm and all, if they want. But most striking is Baker and Bellis's description of the effect of orgasm on sperm retention. As noted earlier, when a woman has an orgasm soon after a man, she literally sucks sperm into her uterus. If she has no orgasm at all, or has an orgasm before the man ejaculates, she retains less sperm. They also discovered that if she orgasms between episodes of sexual intercourse with a man, by masturbating, orgasming in her dreams, or having oral sex with a man or another woman, she also retains less sperm the next time around. The noncopulatory orgasm makes the cervix more acidic and more hostile to the next sperm to arrive in the vagina. Baker and Bellis's research indicates that women *do* have the potential, consciously or not, to alter how much sperm is allowed into their reproductive tracts. This may or may not have an effect on conception—that is, it may or may not have an evolutionary effect—but it surely has an effect on how much sperm of any given male has a chance at all.

More to the point, their study might explain why women evolved orgasm—to control whose sperm goes where, especially when women are dominated by male sexuality. This information suggests that neither male nor female mating strategies rule. Instead, it's a continual arms race to get what one wants.

A Myth of One's Own

My students like to hear about the myths and beliefs of aboriginal cultures. One of their favorite stories concerns the Tiwi of Australia, who see no connection between sex and conception. Before European contact, the Tiwi believed that when someone died, his or her spirit entered into the body of a living female, ready to be born again.[84] Under this scheme, sex, and men, had nothing at all to do with making babies, and this was often confirmed when a woman produced a child even though her husband was long gone on a hunting trip. My students see their belief system as an amusing anecdote about a primitive culture that didn't yet have the advantages of modern science. And yet we have perpetuated a conception myth of our own in Western culture, one that influences the way we think about male-female relationships in general.

THE NEW CONCEPTION

Everyone knows—or thinks they know—how babies are made. Fertilization begins with fast-moving sperm swimming bravely up the vaginal canal into the uterus, and finally into the fallopian tubes. There, sperm encounter a passive egg, lazily making its way down the tube toward the uterus. An energetic sperm attacks the egg and bores inside, penetrating the egg in the same way that the penis earlier penetrated the vagina. This is the conception myth, or scientific truth, we all grew up with. In the last few years, however, several papers have appeared in scientific journals that have the potential to overturn the way we think about conception, female sexuality, and maybe even the interactions of men and women.

Women, and their eggs, aren't as biologically passive as traditional stories of science have led us to believe.

Sperm, as two scientists with the National Institutes of Health pointed out in 1991, have been designed to not stick to tissue surfaces—not the seminal vesicles in men, not the vaginal wall, and not the crypts of the uterine lining.[85] Once in the fallopian tubes, they asked, why would sperm suddenly shift tactics and deliberately head straight for the tangled outer layer of an egg? They discovered that our traditional view of brave penetrating sperm and the patient female egg is basically false. The egg, it turns out, is actually responsible for drawing the sperm in its direction. Although some sperm can be found close to the egg soon after ejaculation, most sperm actually lie motionless in the crypts of the uterine lining, waiting for ovulation. When ovulation happens, sperm wake up from their torpor and once again head for the egg. What makes them revive and move forward? An attracting enzyme released from the follicle of a mature egg during ovulation, it turns out, is the chemical that sucks the sperm in the egg's direction. Sperm are pulled reluctantly into the outer layer of the egg, the corona radiata, and it is only this contact that makes the sperm resume an active role and begin to bore in. The tale of reluctant sperm explains why out of approximately 2 million sperm released by a man during sex, only about two hundred ever get close to the egg. Without the pull of follicular fluid, few sperm get there on their own. The scientists point to a kind of "sperm-egg communication" to explain how conception really works. Although fertilization can't occur without the boring motion of sperm, the pull of the egg is just as essential in the process of conception.

This information is startling for several reasons. On a practical level, it might help explain infertility in couples who have normal sperm, eggs, and reproductive organs. It is possible that a woman unable to conceive is missing that vital enzyme. Once biologists identify the properties of the enzyme, there might be a way to replace the missing enzyme, thereby offering such women a chance to conceive a biological child. But what interests me is the social implications of this discovery. The new data suggest that all our ideas of the male-dominated process of fertilization (i.e. the male himself) are specious. The egg (and the woman herself) is now seen as an active participant in conception. We no longer have to see the process as involving one proactive, aggressive partner (the sperm) and its correspondingly passive and reluctant partner (the egg). It seems reasonable to point out the ramifications of this new model to the way we stereotype male and female sexual behavior. Perhaps now there's more room for the reluctant male and a more assertive female, and a cooperative rather than antagonistic or unbalanced relationship between the two principal actors in the conception drama.

The Post-Modern Sexual Woman

At this point, I'd like to propose what might seem to some like a radical picture of the nature of female human sexuality. I never did believe in all that talk about human pair-bonds and the seductive power of women over men (see Chapter One). Although the possibility of sex might bring two people together in the first place, sex alone never holds a relationship together over time. Also, I was never convinced by the idea that women are naturally less interested in sex than men. This view is more a product of a culture that wants women to be passive and unsexual than a statement about the natural biology of sexually free women. What I suggest instead is that over our evolutionary history, women have been selected not for monogamy, but to be promiscuous. That might seem outrageous, but the signs are there in our biology if we just care to look.

Science today tells us the following about the nature of female sexuality: eggs aren't passive participants in the conception game; menstruation may have evolved to keep the bacteria from the sperm of males out of the reproductive tract; women have the potential for serial or multiple orgasms; women are aroused just as easily as men given the right stimulus; women tend to be more interested in sex around ovulation; ovulation is more regular with regular sex; and women might have the ability to regulate the passage of sperm in their reproductive tracts. What does all this new data seem to suggest about the natural sexuality of women? Some anthropologists are married to the idea that women lured men into pair-bonded relationships by offering them sex on demand. Yet if we look at this scenario from the female point of view, it plays out quite differently. How so? Think about these facts again in light of my hypothesis that women have been naturally selected to be promiscuous. Women may have kept a man around initially to keep her cycle even; if he didn't satisfy her ability to have single or multiple orgasms, she could have chosen a new male. The biological dangers involved in choosing a new partner were decreased by the fact that menstruation protects her body from infections that might arise from switching partners. She can also manipulate sperm to improve the chance of conception with a particular male, or maybe reject unfavorable sperm. The ability to move from male to male certainly improves a female's chance of finding the best genes for her future offspring, or perhaps allows her to make sure her reproductive tract is full of sperm during ovulation.[86] In any case, the composite picture is one of a female who seeks sexual pleasure and has developed biological defenses that protect her from the consequences of an active sex life, a creature designed by natural selection to improve her reproductive success by carefully directing her intimate life.

If we look at various human cultures today, most women aren't what anyone would call promiscuous. Does this observation discount my hypothesis about the nature of female sexuality? Not really. By definition, no woman or man can fully express their natural sexual tendencies, because sex, by definition, requires a partner to be fully satisfying. And that partner always has reproductive and sexual interests of his or her own. Women and men are never really left to their own devices in a sexual sense. Instead, they must compromise, and sexual compromise, too, has been fashioned by evolution so that genes can be passed on. Once again, the dependent human infant requires that women compromise their natural sexual tendencies and stay within a shaky pair-bond.

And so we turn to the other side of the pair-bond, men and the compromises they make, to understand the full story of human sexuality.

Men at Work

I'm always a little nervous about spending social time on the campus where I teach because running into students during off-hours is a little disconcerting for me, and for them. It's hard enough to develop a sense of respectability as a professor—especially when you crack jokes during lectures—but maintaining that professor-student distance during time away from the classroom is even more difficult. And I know from strained conversations with students that they are un-settled as well when we unexpectedly encounter each other at parties, or late at night in the grocery store. But one weeknight last spring, against my better judgment, Tim, my significant other, and I decided to catch a screening of the movie *My Own Private Idaho* at the campus theater. As we took our seats in the large lecture hall full of four hundred or so students, I felt several pairs of eyes on me, checking out their professor on a real live date. No problem, I thought to myself, there's no reason to be self-conscious. I have a life outside the campus, and who cares what students think. The lights went down and the film rolled.

The movie opens with a young man, the actor River Phoenix, receiving oral sex from another man. This "real" scene shifts back and forth to what the boy is thinking about during this sexual interaction—and he's thinking about a house. At the moment of orgasm, the house flies through the air and lands with a monumental, but silent, crash in the middle of a deserted highway. And at that moment of dramatic silence, Tim screamed, really screamed, a yelp so loud that even the popcorn seller in the foyer must have heard him. Several heads turned in our direction. I almost died from embarrassment. There I was, trying hard to be unobtrusive, and only minutes into the film Tim was calling every pair of eyes in the room to turn in our direction—right in the middle of a sex scene. Had he lost his mind? I wanted to kill him.

When the movie was over, I couldn't wait to ask him what that scream was all about. "Oh," he answered, laughing. "Don't you understand how great that falling house was? It was the perfect image of an orgasm." Tim is an artist, and "perfect images" are his forte, so I began to think about why a falling house was such a powerful image of orgasm to him when it didn't do much for me. I didn't really understand why the director had used a crashing house at all. And then it hit me—I didn't scream because while the crashing house might be an appropriate image of a male orgasm, at least for Tim, it doesn't fit any representation that I know of female orgasm.

I tell this story as an apology. The crashing house, and how Tim and I reacted differently, demonstrates a fundamental difference in perception between men and women about sex. And yet I, a woman, presume to write a chapter about male sexuality. And so I decided to start this chapter by admitting that I don't perceive sex exactly like Tim, or any man, and I guess the reader, especially male readers, will have to factor that into the information and interpretations presented here. For centuries, men have written about women in love, in bed, and about female sexuality. In that sense, perhaps my attempt to reverse those roles is not so surprising. I do question and listen to men about their sexuality. And

I've certainly read and studied about male reproductive biology and sexuality. Nonetheless, I feel a bit self-conscious about writing on the intimate details of male sexuality when I don't have a penis and I never felt an orgasm as a crashing house. In any case, my main goal in this book has been to tackle the overall question of the evolution of human sexuality, and so even without testicles of my own, I have to understand mating, reproduction, and sexuality from the male side too.

I'm particularly interested in how science has added to our attempt at understanding why men have sex, and with whom. As with studies of women, most studies of male sexuality that add to our understanding don't deal with human *behavior,* which is almost impossible to track and make sense of, but concentrate instead on what male biology tells us about men.

Enjoy Being a Guy

In our modern cultural times, it is important to make the distinction between sex and gender. Gender is the culturally laden, and culturally defined, term that is imposed upon each of us. Gender is something that is molded by society and can be changed.[1] With regard to gender, maleness and femaleness is defined by what we wear, by our jobs, and by our household tasks. For example, only women in Western culture wear skirts, and this is considered "female." But in many other cultures, such as Balinese or New Guinea, men wear skirt-like garments and this is considered part of maleness. In America, farmers are mostly men, but in many aboriginal societies, women do all the planting and harvesting. Maleness and femaleness at this level are defined by societal consensus and tradition, and biology often has little to do with these decisions. Sex, on the other hand, is one's biological designation as either a male or female, something that cannot be changed except with radical surgery. A person's sex, in the biological sense, is defined by the products of reproduction, and these products are manufactured under the guidance of certain hormones that are driven by a particular chromosomal configuration. It's like a stack of dominoes. Men have a Y chromosome for one of their sex chromosomes, which in turn directs a fetus, early in gestation, to manufacture an antigen that begins the development of testes. The Y chromosome also directs the fetus to produce an inhibitor that stops the development of female reproductive organs.[2] As I explained in Chapter One, the basic human blueprint is female, and so these chromosomally based changes must occur for a fetus to stop its development toward femaleness and redirect it toward maleness. Once developed, the testes, even in fetal life, produce large amounts of the hormone testosterone, and testosterone molds the external appearance of male internal reproductive organs and external genitalia.

THE ALL-PURPOSE HORMONE

Maleness is a complex of characteristics that include large amounts of androgen hormones, a penis and bilateral scrotum as external sexual organs, and the internal plumbing for sperm manufacture. The principal androgens for maleness are testosterone (T), dihydrotestosterone (DHT), and antrostenedione, but testosterone is the most important. It's produced mainly in the testes, but as in women, small amounts are also secreted by the adrenal glands on top of the kidneys.[3] Men have other hormones that direct their sexuality, such as estradiol, but even most of these derive from androgens that are chemically transformed as part of the body's metabolic action.

At puberty, the testicles of young boys begin to pump out large quantities of testosterone. Every physiological change that occurs to boys during this time can be traced, in some sense, to testosterone. It causes the penis to elongate and thicken and the testicles to grow larger toward adult size. Testosterone is responsible for the appearance of facial and pubic hair, the enlargement of the larynx, which deepens the voice, and the shape of a man's body, with wide shoulders and narrow hips. It even has something to do with the sudden production of strong-smelling sweat. In the beginning of the adolescent growth spurt when boys seem to shoot up in height almost overnight, testosterone levels are low. After several years of pubertal growth, high levels of testosterone cause the epiphyses at the end of long bones to seal shut, which has a countereffect and stops growth forever.[4]

Even when puberty has run its course, testosterone continues to play an important role in male sexuality. Although no one knows the exact level of testosterone required for normal sexual functioning, it's clear that testosterone must be present to arouse sexual desire in a male. There's also a relationship between the time it takes to get an erection and testosterone level, but testosterone doesn't necessarily affect how stiff that erection becomes.[5] Beyond sexual desire, the presence of testosterone is imperative for sperm production and delivery.

Testosterone is a tricky hormone to measure because not all of it is free-floating in the blood stream. Much of it, about 30 to 60 percent, is attached to a globulin, called sex hormone binding globulin (SHBG), and more is also loosely bound to the protein albumin. Only about 25 percent of total testosterone travels unbound through the bloodstream and can be measured by blood assay.[6] It's also a highly labile hormone that rises and falls over the course of a day, a week, and a lifetime.[7] Scientists do know that testosterone production begins during fetal life, increases tremendously during puberty, levels off, and then declines with aging. Also, with age more testosterone binds with sex binding globulin, which leaves even less free to be biologically active.[8] The decrease in circulating testosterone probably accounts for the lower sex drive and longer refractory periods between orgasms that happen to men as they age.

Androgens in general affect the limbic system of the brain, especially the anterior hypothalamus, and they are needed for spinal cord reactions including penile erection.[9] When androgens are, for some reason, lost, men lose interest in sex in a few weeks and they eventually can't orgasm. This process can be reversed with artificial androgen replacement therapy.

THE MALE MAP

Culturally, we define men very simply by the appearance of a penis, but the male reproductive and sexual system is actually very complex. As with all mammals, the sex and reproductive organs of men are entwined with the urinary system, which means that the penis is not just a way to disperse sperm but is also used to excrete urine.[10] Because we are terrestrial vertebrates, fertilization is internal, which means the sperm have to get close to the egg somehow. Evolution has opted for the penis as a sperm delivery organ. As in women, sex and reproduction are physically connected. In a sense, the relationship between sex and reproduction is even closer for men because their gametes, sperm, are almost always part of sex, while the female gametes, eggs, are only occasionally mixed up in a sexual interaction.

Sperm are manufactured in the testes, balls of convoluted seminiferous tubes set in scrotal sacs that swing free under the penis (see below for a more detailed discussion of sperm production). Testicles develop high in the abdomen during fetal life, but unlike what happens in most other animals, where the testicles remain in the abdomen, they descend into scrotal sacs outside the abdominal wall soon after birth. Even when testes do descend in some other species, the migration usually doesn't happen until sexual maturity. This is why it's sometimes difficult to tell if kittens are male or female. No one is sure why some mammal testicles in general, and primates' in particular, are compelled to migrate down and out. Some have assumed that sperm production requires a cool atmosphere to work correctly. This explanation doesn't make complete sense, however, since many animals with testicles high in the body do just fine. Birds, for example, have high body heat but their testicles are still internal. It might be that sperm storage works best at cool temperatures and so the sperm storage organ, the epididymis which is attached to the testes, must be exposed to cooler air.[11] We do know that for humans in particular, testicles must be descended and free for sperm production to take place at all. Human testes are larger than other primates' and they hang lower, although chimpanzee males also have large swinging sacs.

Once manufactured, sperm are dumped into the epididymis, a storage and maturation area of ducts drooping like a pile of wet pasta over each testicle. The epididymis is a highly convoluted series of ducts, about 20 meters long when unfolded. It acts as a sperm reservoir and as a filter; some aging and defective sperm are rejected here and absorbed back into the body. As sperm mature in the epididymis, fluid is added to the mix, and after twelve days or so the sperm move out to the vas deferens.

The vas deferens is two long corridors, each 40 centimeters long, 0.5 millimeter wide, that begin at the testicular epididymises, run up the abdomen, loop over the top of the bladder, and then scale downward to join the urethra below the bladder at the prostate gland. Contractions of smooth muscle move sperm along, and this motion increases with sexual excitement.

Close to the bladder are several glands that manufacture liquid that makes up semen, the fluid of ejaculation. The seminal vesicles, bags on either side of the bladder, mostly add fructose, or sugar, to semen. The prostate, a large gland lying between the opening of the bladder and the beginning of the penis, also adds more seminal fluid. And finally, two other glands add to the seminal mix, the Cowper's glands, two small beads on either side of the penile base, and the urethral or Littré's gland in the body of the penis.[12] The liquid mix from these glands nourishes the sperm, protects them, helps them glide though the male reproductive tract, and provides a medium for transport and protection once ejaculated. These chemicals also help the sperm in their next life stage by causing the walls of the female reproductive tract to contract, which of course helps the sperm move forward. Semen also provides nutrients for sperm movement and survival outside the male tract. A human ejaculation is typically 3 millimeters of this fluid, about a tablespoon full, with a relatively low density of sperm. Even so, men usually ejaculate about 250 million sperm with each orgasm.[13]

As I mentioned earlier, the human penis can be distinguished from other mammals' by the lack of a bacculum, or *os penis,* a small bone that appears in other primates but was selected against in our ancestors for some reason no one has yet figured out. Instead, the human penis is made up of three rods of tissue without skeletal support. Two are called corpora cavernosa, the same tissue that makes up the body of the clitoris. These fused columns lie parallel to each other and form the body of the penis. They separate at the base of the penis and attach to the front of the pelvic arch by way of the ischiocavernosa muscles. The corpora cavernosa are made up of sponge-like tissue that fills with blood and makes the penis erect during excitement. Underneath the two corpora cavernosa is another shaft of tissue, the corpus spongiosum, which doesn't have quite the erectile properties of cavernosa although it still fills with blood and stiffens. The weaker stiffness of the spongiosum makes sense because the urethra runs through it and sperm need to flow down an open urethra during ejaculation. The spongiosum expands at the tip and forms the glans of the penis. On the outside, the glans can be distinguished from the shaft of the penis by a rim which give the glans a helmet-like appearance, especially if a man is circumcised or the foreskin is pulled back. The foreskin is attached to the underside of the ridge of the glans and it retracts during sexual excitement because the skin of the penis is pulled taut.[14] The corpus spongiosum expands at the other end to form the urethral bulb, where semen build up right before ejaculation, and it's attached to the pelvis by another pelvic muscle, the bulbospongiosus. All three shafts are surrounded by a tough fibrous lining called the tunica albriginea, which helps makes the penis rigid when erect.[15] The penis is supplied with blood by a branch of the pudendal artery, which then divides into two major branches and sends blood up the dorsal side as well as through the middle of the penile body. The penis has the same number of nerve endings as the clitoris, but obviously these nerves cover a larger area.[16] Testosterone makes the penis highly sensitive to tactile sensations, which are transferred by the pudendal nerve to the base of the spine.

MALE ORGASM

Masters and Johnson pointed out years ago that most societies fixate on the male erection.[17] It's odd, they further commented, that our culture worries about women who don't orgasm, but focuses on men who don't get an erection, rather than worrying about male orgasm too. Along those same lines, impotence is defined as the inability to get an erection, not the inability to orgasm or ejaculate, and aphrodisiacs for men are potions and lotions to help get or maintain an erection. And yet orgasm, not an erection, is the real pleasure of sex—and the real point, in evolutionary terms, behind the penis delivery system.

For males, orgasm almost always involves ejaculation, the expulsion of semen through the urethra. Expulsion can also occur without orgasm, such as in nonorgasmic nighttime emissions, but most often an orgasm involves sexual excitement, erection, and ejaculation. Men undergo the same sexual phases outlined by Masters and Johnson as women—excitement, plateau, orgasm, and resolution.[18] In the excitement phase, penile erection can come and go without much urgency. As I explained in Chapter Two, penile erection, vasocongestion of the pelvic area, and body myotonia, define the excitement phase and occur as a primary response to sexual stimulation. That stimulation isn't necessarily mediated by the brain. The penis becomes erect as blood flows into the corpora cavernosa and corpus spongiosum and is captured when arteries restrict blood from leaving the tissue. No one is quite sure about the exact hydrology of erection except that blood comes in and is not let out until orgasm, or until the sexual stimulation has ended. During excitement, the testes also swell and are elevated toward the body as the dartos and cremaster muscles supporting the scrotum contract. The walls of the scrotum thicken and contract as well. In addition, male nipples become erect under sexual stimulation and the body tenses.[19]

During the male orgasmic plateau phase vasocongestion continues. In particular, the glans penis swells and becomes darker. Some fluid, mostly excretions from the Cowper's glands, can leak down the penis and out the urethral opening at this point. This fluid might contain a few sperm, but mostly it is leakage of spermless fluid.

The orgasm stage is divided into two levels by Masters and Johnson.[20] In Stage I, the smooth muscles of the sperm-holding organs, the epididymis and the vas deferens, begin to contract and push sperm along in seminal fluids toward the prostate. The first three or four contractions are the strongest. At the same time, the glands that contribute seminal fluid also contract.[21] Fructose-rich fluid pours out of the seminal vesicle into the prostate which contracts rhythmically and passes along the collected seminal fluid to the urethral bulb, where more contributions are added from the Cowper's glands. In a sense, semen at this point is like acrylic paint. Sperm is the pigment, the element in semen that gives it purpose in life. But color, or sperm for that matter, can't be delivered and expressed without the mover, which is the seminal fluid that carries the pigment. The end of this stage occurs when the urethral bulb expands two to three times its normal size, exerting pressure on the pelvic area. During this rush, the sphincter to the bladder closes off and a man feels that orgasm is imminent.

In Stage II of orgasm, the pent-up fluid is released from the urethral bulb and propelled along the penile urethra by rhythmic contractions of the urethral bulb and the muscles of the penis, the bulbospongiosus and the ischiocavernosus. The entire length of the penile urethra also contracts, propelling semen along its length. This part of Stage II is ejaculation, or release of seminal fluid down the penile shaft. The other part of Stage II is the orgasm itself, which involves contractions of all the muscles of the pelvis and the accompanying mental feeling of release.[22] These contractions take place one to six seconds apart and there's complete relaxation in between. Men average about seventeen contractions per orgasm. Interestingly, each man has a particular contraction pattern that identifies him like a fingerprint.[23] During orgasm a man also has the same heart and respiratory increases as a woman, and sometimes the same orgasmic flush on his skin, and the same sort of myotonia, or muscle tightening and release.

Resolution, the fourth stage of male sexual response, occurs after orgasm and ejaculation are complete. Nerves and muscles so previously responsive to sexual stimulation are now physically unable to respond. This might last for a few minutes in young men and up to an hour in older men. Fifty percent of the erection is lost immediately after ejaculation and then the penis slowly returns to the flaccid state. The time in between erections is called the refractory period.

The above tells us how men have sex, but it really doesn't say why. Certainly men have sex because it feels good, but there's a more important reason sex is connected to pleasure. Over millions of years, natural selection has linked orgasmic pleasure to delivering sperm as a way to encourage men to have sex. And so spreading sperm, and passing on genes, is dependent on the rewards of human sexual satisfaction.

Gambling with Sperm

E very animal organism has been preprogrammed to pass on genes. Some individuals are better at this than others, but we all have the hard wiring that encourages us to expend energy on mating and rearing offspring. Evolutionary theorists call this our "reproductive effort," the amount of energy that each of us devotes to finding mates, having sex, and bringing up offspring. Each sex is expected to allocate their reproductive effort in different ways. Females, who by design become pregnant, and being mammals must lactate infants, supposedly expend most of their reproductive effort not in finding mates, but in parenting. Males, on the other hand, are most often biologically designed to throw their primary effort into mating rather than parenting, although human males are under a certain amount of pressure to invest in infants too. Males play the mating game with sperm, those tiny mobile gametes that are a male's pack of renewable cards to be tossed into the game of reproductive success.

SPERM STORY

Philosophers and naturalists used to think that each sperm carried an exact replica of a whole man in the head, a homunculus or tiny man enclosed in the sperm head. In the late 1800s, biologists used the light microscope to discover the head was actually all nucleoprotein, the male's genome. By the turn of the century, scientists looking at external fertilizers such as fish finally understood the basic properties and conceptive function of sperm. Researchers also observed that sperm cells seemed to stay active for hours under the microscope and could be transported if kept at a constant low temperature in a thermos. Soon after, sperm began to travel. The first record of artificial insemination involved a cross between a Polish ewe and a Suffolk ram in the early 1900s, which produced an Anglo-Polish lamb.[24] Artificial insemination and domestic breeding were enhanced with the invention of the artificial vagina in the early part of this century; animal breeders could then direct the evolution of their herds by manipulating mating and fertilization. The same technology is used today when infertile couples bypass sex and use *in vitro* fertilization, where sperm and eggs are joined in a petri dish and then reinserted into the uterus, or when women conceive with sperm from men stored at a sperm bank. New technology even allows semen to be fractionated into X- and Y-bearing sperm to increase a couple's chances of having one sex infant over another. All this to manipulate the tiny cells constantly dividing in all men of the species, even as you read.

THE PRODUCTION LINE

Testicles have three functions. They produce sperm, they're responsible for most of the production of testosterone in males, and they provide the primary fluid, a plasma, for developing sperm. About 30 percent of the testes is taken up by Leydig cells, which secrete about 10 milligrams of testosterone a day.[25] This testosterone, as I mentioned earlier, is mostly bound to protein, but some is free-floating in the bloodstream. It appears to affect almost every aspect of what we think of as "maleness," especially the appearance of secondary sexual characteristics such as beard growth. It also influences the accessory organs of male reproduction such as the prostate and Cowper's glands. And it's especially important for the initiation of spermatogenesis—the manufacture of sperm which begins at puberty and continues throughout a man's life.[26]

The sperm of every species is so different in shape and size that reproductive biologists could use the differences to make up taxonomic charts. Human sperm is about 58.4 microns long, with a head which carries half the complement of DNA (so that it can be combined with an egg which also carries half), a midpiece, and a swirling tail.[27] Sperm production begins at puberty and eventually reaches an output of 500 million sperm per day.[28] The properties of sperm that influence fertilization include not only overall sperm count per ejaculate, but the motility and morphology of the sperm.[29] Morphology refers to the shape of the head and tail; obviously deformed sperm don't travel as far as normal

Most cities relegate the commercial side of sex to a certain district, but these districts are also a vital part of the city where average people live and work. (© DEDE HATCH)

Fossil remains of our ancient ancestors the Australopithecines show that three and a half million years ago males were much larger than females. This dimorphism in size is usually associated with a polygamous mating system. Male and female humans today are much closer in size.

(PHOTO. FINNIN/CHESEK. COURTESY DEPARTMENT LIBRARY SERVICES, AMERI-CAN MUSEUM OF NATURAL HISTORY)

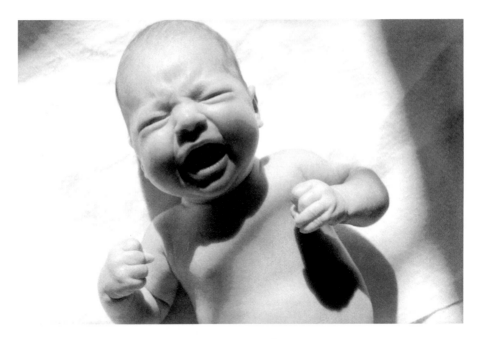

All primates are distinguished by periods of long infant dependency. Human infants usually require a heavy investment of time and energy by both the mother and father. (© DEDE HATCH)

Anthropologists have suggested that infant dependency has selected for the human pair-bond, a system whereby couples cooperate to bring up offspring. (MEREDITH SMALL)

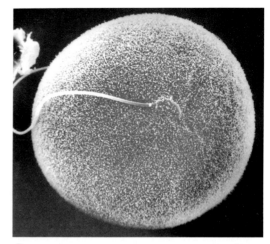

Recent research shows that old notions of courageous sperm battling to impregnate a passive egg are inaccurate. The egg, it appears, plays a much more active role, emitting an enzyme that pulls the reluctant sperm toward it. (PHOTO RESEARCHERS, INC. © D.W. FAWCETT)

Humans may have been naturally selected for pair-bonding, but they haven't been selected to be sexually monogamous. In fact, sexually faithful couples are as rare today as they probably were in our ancient past. (© DEDE HATCH)

Gender roles are culturally, rather than biologically, prescribed. In the Woodabe tribe of Nigeria, men must dress extravagantly, and elegantly, to attract the favors of women. (© CAROL BECKWITH)

Recent studies on the biology of male homosexuality have investigated genes, hormones, and brain biology for clues to human sexual orientation. (© DEDE HATCH)

New research has shown that when couples are apart for a significant period of time, men release much larger numbers of sperm the next time they have sex with their partner. (© DEDE HATCH)

Evolutionary theory predicts that men should desire young, healthy mates with long reproductive careers, while women should seek out men with generous resources to share. But in practice, humans tend to marry those of the same age, socioeconomic class, religion, and race. (© DEDE HATCH)

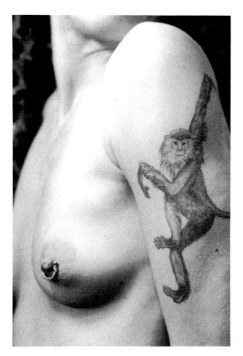

Just about anything can be a sexual stimulus for humans. Touch, sight, smell, and sound combine to arouse us. (© DEDE HATCH)

Several studies have shown that when women spend time together in close contact, they eventually synchronize their menstrual cycles. (© DEDE HATCH)

Females in about 10 percent of primate species display flamboyant signs of estrus, such as this perineal swelling on a Barbary macaque. Human females effectively conceal ovulation. Some suggest this concealment has had a major effect on the evolution of human mating patterns. (MEREDITH SMALL)

At the deepest level, we are driven to attract mates and have sex because we are driven by our inner biology to pass on genes. (© DEDE HATCH)

Less science has been directed toward understanding female homosexuality. Our Western culture, it appears, is less threatened by lesbian love than gay love. (© DEDE HATCH)

The Internet and other electronic inventions are opening new, less intimate, avenues for human sexuality. (© DEDE HATCH)

Sex in the future will continue to be influenced by a combination of culture and biology as humans continue to evolve. (© DEDE HATCH)

sperm and are unable to enter an egg. Motility, rather than mere sperm count, is the best predictor of fertilizing ability; a man with a few fast-moving sperm is more fertile than a man with billions of slow movers.

Within the testicles is a series of tubes called seminiferous tubules which are 0.4 millimeter across at their widest. These tubules end in the rete testes, an area that functions as a launchpad for sperm on their way out of the testes. There are two cell types in the layers of the seminiferous tubules, supporting cells called Steroli cells and a physically active group of germ cells that multiply and divide and eventually become sperm.[30] Spermatogenesis, or cell division of the germ cells into functional sperm, begins as these cells first divide into two daughter cells. The result is two cells with a full number of chromosomes, as if skin or blood cells had duplicated and then divided. The two cells then go through another division, dividing their full set of chromosomes into two equal halves. The result is four sperm cells that contain only half the number of normal chromosomes, including the ones that determine a person's sex. This is where the sex of future offspring get their blueprint. The original germ cell has, of course, both an X and a Y sex chromosome. It first duplicates both those X's and Y's. But when those two daughter cells divide into four, the X's and Y's stand alone in the resulting sperm cells. From the original germ cells arise four sperm, two carrying the female-determining X chromosome and two bearing the male-determining Y. When one of those sperm eventually joins with the female egg, which always carries an X chromosome, the sex of the child is fixed; if an X sperm joins an egg, the conception is female, and if a Y sperm fertilizes, the fetus is male. In other words, the sperm of men, not the eggs of women, actually determine the sex of offspring.

Just because the sperm production line is ever moving doesn't mean this is a fast process. From germ cell to sperm takes about two months in the testicles, and sperm still have a long journey ahead of them. At least they don't have to start on this journey alone. Groups of germ cells tend to divide in a coordinated fashion, which produces a cohort of immature sperm. These cells look like regular sperm but they are virtually immobile and unable to propel themselves out of the testicles. Instead, the undulating action of the seminiferous tubules internally pushes them through the rete testes area and eventually into the epididymis, where they mature, move on to the vas deferens, and then wait for orgasm and ejaculation.

THE DANGEROUS JOURNEY

Reproductive physiologists agree on one important point—no one really knows exactly what happens to sperm inside the female reproductive tract. The first piece of bad luck for human sperm is that they're deposited in the vagina rather than directly into the uterus as they are in horses, pigs, and cows. Once ejaculated, the semen forms a coagulated plug that stops sperm action for a few minutes until it melts. Once free from the melting plug, awake and ready to swim, the sperm must first deal with a barrier of cervical mucus that drapes across the opening of the uterus like a liquid curtain. Most of the time, cervical mucus is formed of interlocking mucoproteins which prevent sperm from entering the uterus; the ejaculate, and all the sperm, usually ends up dripping out once a woman stands up. But under the influence of estrogen during the week surrounding ovulation, cervical mucus becomes more stretched out, more viscous, and more friendly to swimming sperm (see Chapter Three). If the conditions are right, sperm then pass through the cervix and enter the uterus within minutes after ejaculation. Cervical mucus is also a supportive medium at this time, providing a suspensory agency that moves sperm into the uterus over several hours or days.[31]

At this point sperm undergo a process called capacitation. The head of a sperm is covered by an acrosomal cap which is wiped off as the sperm moves through the female organs. Like taking off a hat when entering a house, this process essentially prepares the sperm for entering an ovum. But first the sperm must swim, and be pushed by waving cilia and subtle contractions of the uterine walls, toward the junction of the fallopian tubes.[32] In a sense, the uterine atmosphere is hostile to sperm. It's acid in character and full of nooks and crannies that detour sperm from moving closer to an egg. Although as many as 250 million sperm might be deposited at the cervix, fewer than 200 avoid being caught in cervical mucus or the crypts of the uterus and end up in the fallopian tubes. Some reach there within thirty minutes of leaving the penis, while others show up days later. And they do have survival power. Wiggling sperm have been found alive within the female tract as much as eight days after sex, although no one knows if these sperm could actually find their way to an egg at such an advanced cell age.[33]

From an evolutionary point of view, the sperm are everything—these are the mobile packets of DNA that carry genes to the next generation. But to a man, or a woman for that matter, sperm are just invisible creatures supposedly contained in the milky-white fluid that shows up during sex. Actually, the two views are not incompatible. Sperm essentially use orgasm and ejaculation to gain entrance into the female tract. Without orgasm and ejaculation there would be no sperm deposits, and so sperm are intimately implicated in how men and women go about their sexual matings.

THE ODD CASE OF SPERM

Why males make so many sperm is somewhat of a mystery. It might be that they need to be deposited in large numbers simply because there's so much attrition along the way. If you start out with, say, a few million, and then drag them through a funhouse of twists and turns, at least a few hundred will make it to the end point.[34] Or it might be that the production line makes so many duds that high numbers are necessary just to get any good ones at all. If half the sperm produced are going to be headless or double-tailed, it's better to produce extravagant numbers.[35] And finally, males might have evolved the need to lay down huge number of sperm to beat out the competition in conception.[36] If females mate with more than one male, each male is faced with the possibility that his mating is for naught. But if his sperm are better, higher in numbers, and more vigorous than the other guy's, he might outrun the competition as sperm move toward the egg. Each sperm then becomes a kind of lottery ticket owned by one man, but with a different number tattooed on the tail. Given the laws of probability, even when other men have tickets, the one with the most tickets has the higher chance of winning the conception game.

In any case, there are certainly more sperm, or male gametes, around than there are female gametes, or eggs. But this doesn't mean that sperm are always available, always good for fertilization, or free. Biologist Donald Dewsbury was the first person to evaluate the cost of sperm production.[37] As Dewsbury pointed out, men might make more gametes than women make eggs, but they don't release those sperm one by one. Instead, mating by a male involves the deposit of a bundle of millions of sperm in one shot, and this bundle has its costs. Energy is required to move the production line and keep it in working order. No one knows exactly how much energy it takes for sperm production, but it must take some caloric output for cell division, production of semen liquids, and the movement of sperm through the male reproductive tract. Also, every sperm deposit is different; they vary in sperm numbers and quality of sperm for each male over time. And so comparisons should be made not at the level of sperm and egg, but whole ejaculates to one egg. Given that comparison, women have it easier because they only mature one egg at a time, and over long intervals, while men are continuously in production. Although each sperm is theoretically potentially capable of conception, many are lost in the process. More important from the cost side of the equation, a man can only deposit so many of those packets each day. In all males, there's a certain turnaround time after ejaculation before another bundle of sperm is gathered for deposit. In other words, no man really has unlimited sperm supplies or the ability to copulate and deposit sperm willy-nilly.

COMPARATIVE SPERM

The sperm of human males, it seems, don't fare well when compared to most other primate males'; pound for pound we're commensurate with gorillas, but we pale in comparison to chimpanzees. Big animals tend to have bigger testes; apes, for example, have bigger testes than monkeys; humans, too, have relatively big testes. In general, the larger the testes, the higher the sperm count per average ejaculation.[38] This is simply a matter of more testicular tissue and a higher mass of cells for gametogenesis.[39] But the size of testes relative to body weight is important as well. Testicles in humans account for about 0.08 percent of the average male body weight; in our close relative, the chimpanzee, testicles account for about 0.30 percent of body weight, about four times as much. Both chimpanzees and gorillas fire off ejaculates that are more densely filled with sperm than human ejaculate.[40] Daily human sperm production is estimated at 185 million sperm per day, based on the combined average testicle weight of 42 grams and how many sperm that much testicular tissue can produce.[41] Most men ejaculate 3 millimeters of semen, which contains 150 to 360 million total sperm. About 40 percent of those sperm are abnormal and unable to successfully navigate the female reproductive tract or manage fertilization.[42] And the fact that a male has not ejaculated in several days doesn't mean there are great reservoirs of sperm waiting to be used. Some sperm is stored outside the testicles in the epididymis and in the vas deferens—about 225 million—about the same number used in a single average ejaculate.[43] In other words, the average male only has about one extra ejaculation waiting in the wings. Compare this to a male domestic sheep that has ninety-five packets in waiting.[44] Stepping up production doesn't help either. Excess sperm beyond what the male body is biologically designed to hold is just resorbed into the body and its place is eventually taken up by fresher sperm cells. Other mammals, including most primates, make more sperm per day, store more sperm in their tracts, and ejaculate much more sperm each time than we do. For example, the average sperm count per ejaculation for a human male is 60 million sperm per millimeter of semen; in contrast, a rhesus monkey male releases 1 billion sperm per millimeter. Monkey testicles also manufacture sperm at a rate six times higher than human testicles.[45] As a result, if a man ejaculates, on average, 185 million sperm each time he has sex, more than one orgasm a day would put a strain on sperm supplies, and secondary ejaculates would be sperm-light.[46] In fact, a man's sperm count drops by 72 percent if he ejaculates more than once a day, and having sex more than 3.5 times a week significantly decreases total sperm supplies.[47] It's unknown, however, if that decrease actually cuts into a man's ability to fertilize a ripe egg.

This comparative sperm information is important because it tells us something about our "natural" mating system. Theorists think the relationship is rather straightforward—when males are faced with the possibility of setting their sperm alongside the deposits of other males, they will produce huge numbers to outdistance the competition. And the numbers bear this out. Several researchers gathered information on sperm counts and measurements of sperm quality, such as motility and average percentage of abnormal sperm, over the primate order and compared these numbers with the species-typical mating system as seen from the male's point of view. Males who mate within a system that includes other males, that is multi-male groups like baboons, chimpanzees, and macaques, have large testes and send off high-quality, extremely motile sperm. But when males mate with one female in a monogamous system, as gibbons do, or with several females in a harem all his own, as gorillas do, sperm quantity and quality are low.[48] These numbers are important because sperm reflect the mating tactics that males and females have evolved. Females mate a certain way because they need sperm to conceive. But females might also incite competition as a tactical move among males looking for a partner.

Human males are no different from other animal males—they evolve mating strategies and sperm types based on what females do. If faced with sperm competition, they would deposit large loads. And without much competitive pressure, they might ease up and put their energies elsewhere. Since human males have low sperm counts and not much backup, require long periods in between to "reload" their sperm supply, and because sperm quality is not very good, men must not be under much selection pressure to compete with sperm from other males that might find their way into females. Based on overall comparative sperm counts among primates, human males must be one of the species adapted to a low level of sexual activity that isn't reproductively threatened much by other males.[49] Maybe.

Male Mating Strategies and Sperm Competition

M ating or conception isn't always a matter of having a steady mate. Even for marginally monogamous primates like ourselves, the chance of cuckoldry is always close at hand. Recent studies of sexuality in Western cultures show both men and women engage regularly in extramarital sex.[50] As many as 50 percent of married men and women now admit to having had an extramarital liaison, although no one is really sure what the exact percentage of the population engaging in extramarital sex might be. The same kind of philandering is true for both sexes in non-Western cultures as well.[51] In every society where men have affairs, at least a few women do, too. Women in 73 percent of cultures around the world admit to at least one extramarital affair, and it's a common occurrence in over half of the societies on which we have information.[52] These data are remarkable, considering women are universally punished for adultery much more often, and more severely, than are men; yet these women still have sex outside marriage.[53] And even the so-called double standard toward sex doesn't work in the real world—promiscuous men must be having sex with some women, and it doesn't seem likely that only a small group of women are engaging in sex with all these sexually motivated men.[54] It must be that women are more promiscuous than they say. If so, men must have evolved strategies to counteract female infidelity.

PASSING ON GENES THE HARD WAY

Although we humans don't mate in random multi-male multi-female groups, we also aren't particularly monogamous either. This presents a special problem for males who need to pass on their genes. Women ovulate so seldom and are pregnant for so long that finding a fertile mate and monopolizing her sexually is not easy. Despite sexual pair-bonding, women are as likely as men to slip off and sexually interact elsewhere. Now, such an affair might be meaningless fun in the evolutionary sense if the woman isn't near ovulation, and presumably many of the extra-pair mating admitted to in our culture and other cultures are of this kind. But there is also the very real risk of pregnancy with a partner other than the committed spouse. In a recent survey, it was estimated that one out of every thousand copulations by committed couples is a "double mating" for the women. This means she has sex with her partner and also with another man within five days.[55] When she does this, the sperm of her long-term partner is set in competition with those of another male. This competition is most significant when the woman is near ovulation and the sperm are allowed though the cervical mucus. And this does happen. Several studies in the United States and Great Britain have shown through blood analysis that about 10 percent of babies couldn't possibly have been fathered by the men named on the birth certificate.[56] In one study where men acquiesced to the legal responsibilities of paternity without undergoing blood testing, 18 percent were shown later not to be the real fathers.[57] The result is that almost half of societies worldwide express a low level of confidence about exactly who is the father of whom.[58]

If this is all true, men are often at risk for cuckoldry. This vulnerability might not involve sperm competition in the sense of chimpanzees, where females are mating with many other males on the same day and in broad daylight in front of the group at large. But it is a level of potential sperm competition nonetheless. More important, our sex acts happen alone, in private, and often in the dark. And so any sperm competition is a secret skirmish among males who probably never see the whites of their competition's eyes. We might expect, therefore, that men have been selected to evolve intricate strategies for combating the tendencies of women to mate elsewhere.

PULLING THE CORK

If men were monkeys or rats, life would be much easier. Mixed into the seminal fluid of many other mammals are clotting enzymes and coagulable proteins that hang together to form a plug within the female vagina. This plug not only helps keep deposited sperm inside the female, but it sometimes helps dissuade the sperm of subsequent males from making much headway in the female reproductive tract. In addition, vaginal plugs serve as within-female storage sites for sperm; as the plug liquefies bit by bit, suspended sperm are time-released into the female reproductive tract.[59] Human semen also coagulates, but it liquefies quite quickly. This presents a special problem for a bipedal creature—mere standing up will rid the female body of a large proportion of sperm no matter who put it there.[60] But there are ways around this dilemma.

One possibility, strangely, comes from looking at the properties of "bad" human sperm. More than other primates', human ejaculates are noted for the high number of malformed, "useless" sperm. Biologists have always assumed that these abnormal bits are a necessary byproduct of the sperm production line.[61] Since males make so many billion sperm, there's bound to be a certain number below par; it's a matter of imprecise factory manufacturing versus personal craftsmanship. Another recent and revolutionary view is that nature has a role for these odd-sperm-out; bad sperm might actually be good sperm. Infertility studies have shown that only about 1 percent of sperm are actually able to fertilize an egg.[62] The rest are just seminal "noise." The question is why Nature would opt for so much noise, especially in humans who are supposed to be reasonably monogamous. Biologists Robin Baker and Mark Bellis have suggested that abnormal sperm could serve a purpose if they link together to form impenetrable barriers. These barriers would form not at the cervical opening as a normal plug does, but up in the female reproductive tract where the fallopian tubes meet the uterus. Built up in the wake of more mobile, better fertilizable sperm that had made it into the tubes, the barriers would essentially close up the area and prevent sperm from other males from gaining access. It's also possible that their action might trigger female lymphocytes, which kill off the sperm—and the sperm of other males as well.[63] These "kamikaze" sperm, as Baker and Bellis call them, may be a human male's answer to the loss of a copulatory plug. Kamikaze sperm might keep viable sperm inside by plugging the female reproductive tract and keeping out the sperm of competitors.

The only problem with this unique idea is proving it. Baker and Bellis developed their hypothesis with coagulating rat semen in mind; no one has yet shown that human sperm actually cling together and form any sort of barrier. Also, not that many potentially kamikaze sperm get very far inside the female tract in the first place; cervical mucus screens out most of the "bad" from the "good" sperm.[64] In one study of human ejaculations, 69 percent of the sperm collected in the female vagina were abnormal in shape, but that proportion of abnormally formed sperm dropped to 40 percent once past the cervical canal.[65] This suggests that even if some sperm were designed as kamikazes, not many would make it to the utero-tubal junction to be of much help. And finally, there are other ways to keep sperm inside women. As I mentioned in Chapter Three, keeping a female sexually satiated and remaining on her back after sex is one strategy that would be an advantage to a male with liquid sperm supplies, and might prove a better sperm saver than a safety net of so-called kamikaze sperm.

USING WHAT YOU'VE GOT

One of the more interesting features of the male ejaculate is how each sample varies from man to man, and from emission to emission. There's even some evidence that sperm counts, and the testicle size that's associated with sperm counts, vary with race.[66] The point is, ejaculates vary and men are not always able to deposit sperm whenever they want. This is a disadvantage, but it also implies that men might be able to unconsciously manipulate their sperm supplies to their own reproductive ends. New data, again from Baker and Bellis, suggest that men have the capacity to alter their sperm deposits to further their reproductive success. No one is suggesting that men do this consciously. It's not as if a man says,

"Hey, this evening's lovemaking might result in passing on genes, so I guess I'll unleash all my sperm," or, "Hum, there's no chance of sperm competition tonight, so I'll release a smaller number." It just means that over generations, men who put the correct volume of deposits in women passed on more genes. In other words, their bodies responded appropriately in a way that furthered their reproductive success. The biggest mystery is how the body knew enough to follow a certain plan.

Baker and Bellis conducted a complex study that involved a lengthy questionnaire about sex, collecting ejaculations during sex and masturbation in twenty-five couples, as well as studying flowback semen gathered by women after sex. As a result they were able to estimate the ejaculated sperm counts of men, how much of that sperm was ejected as flowback by their female partners, and what those numbers meant in the larger context of normal human mating.[67] They discovered that the number of sperm men deposited didn't vary with the cycle stage of their partner or whether or not she was taking oral contraceptives. Unexpectedly, fertility, or more crudely, an unconscious push by men to get women pregnant, doesn't seem to be a driving force. Perhaps men don't deposit more sperm into the vagina near ovulation because they aren't aware that ovulation is near, although this seems like an odd assumption, given that it would be in the best interests of men to have evolved mechanisms to detect ovulation. After all, passing on genes is precisely the point—why not synchronize sperm and egg? Nonetheless, it seems that men, in fact, are clueless as to the timing of ovulation and its potential payoff. Nor do men deposit more sperm as a result of female orgasm. If women are using orgasm to manipulate sperm deposits, as I suggested in the previous chapter, men have no counterstrategy for this. In other words, men, or at least their sperm counts, don't seem to be influenced by their partners and what they do with sperm.

Given this, another strategy males could have adopted would be to always deposit the same amount of sperm, regardless of conditions. But in fact, not all sperm deposits are created equal. Two factors routinely make men release more sperm—female body size and the time a man spends in her company.[68] In the first case, it's possible that the data are the result of selective mate choice; large men with large testicles and high sperm counts tend to mate with large women and leave large deposits. It would also make sense that the general physiological situation compels men to fill large women with more sperm. That is, more sperm end up in large women because big women have larger reproductive organs and lengthier tubes to traverse and men, or their testicles, sense this.

More intriguing is Baker and Bellis's finding that the time a couple spends together has a significant effect on sperm count, which may explain why any single man has a variable record of sperm count over time. Twenty-five couples participated in the phase of the study that focused on sperm counts and the amount of time the couple spent together. All were either married or in committed relationships. Baker and Bellis discovered that men deposited less sperm in their partners if they spent a lot of time together, independent of the number of times they had sex. When the couples spent little time together, the man was apt to deposit large amounts of sperm, even if they frequently had sex. Baker and Bellis suggest that in the first case, men deposited low counts because their female partner was basically always around, and thus the threat of sperm competition was low. But when the couples spent hours or days apart, the man responded by releasing a higher number of sperm; somewhere in his unconscious he knew she had had opportunities to be filled with someone else's sperm. This idea conveniently fits with the proposition that humans aren't particularly monogamous, and the data on extramarital affairs worldwide. If women routinely, or even occasionally, stray and a man might never know about her philandering, he would benefit in evolutionary terms from releasing more sperm during times of vulnerability.

In this way, human males may have been dealing with the vagaries in the human mating system for millions of years. Women don't exhibit extravagant signs of estrus, and keeping one's mate sequestered at all times is not an energy-efficient mating strategy. A better biological way to deal with these potentially wandering women is by seminal subterfuge. Sperm competition, though not fierce among human males, is still part of why we have sex and how men manipulate their gametes to increase their odds at reproductive success.

WHY MEN MASTURBATE

The only crack in this nice hypothesis is solo sex. All men masturbate—young men, single men, married men. If the ultimate purpose of male sexuality is to spread sperm around, it seems that Nature has played a reproductive joke on men by making them also enjoy masturbation. Or so it would seem. But Baker and Bellis have suggested an answer for this, too. Many of the men in their subject couples masturbated in between sex with their partner; the researchers obtained sixty-seven ejaculates from these masturbations. Men were most likely to masturbate if they hadn't had sex with a partner for a while, which seems obvious. But from an evolutionary view, this also seems, on the surface, like a stupid strategy. When a man masturbates, and then has sex with a partner within seventy-two hours, his sperm count is significantly reduced. Why would men, who have been selected to maximize how they pass out sperm, waste their supplies?

It's all in the reception, Baker and Bellis contend. They discovered that the women in their study who had sex with these men at some point after the men masturbated somehow retained the same number of sperm as they did when he had not masturbated within the previous few days. In fact, the women apparently retained even more sperm if the man masturbated in between their sexual intercourse.[69] This seems to indicate that even when sperm supplies are low, women can regulate how much sperm they retain or let out. Given that result, Baker and Bellis feel that from a male's point of view, masturbation isn't a negative, and in fact, it has its biological advantages. Men only hold on to about two full ejaculates at a time; the rest of their sperm is resorbed into the body. If a man goes too many days without sex, there's a chance that his vas deferens and epididymis will be full of aging sperm on their way to destruction and resorption. Masturbation might be a way to clear the system and better prepare for the next session of "real" sex. In this context, masturbation isn't really a sign of heightened male sexuality, sexual interest, or desire, but a strategy selected over time to keep sperm supplies fresh in a species that isn't very sexual.

MALE MATING STRATEGIES

What makes Baker and Bellis's work so remarkable is the possibility that male sexuality is geared toward dealing with two factors over which males have no control—female behavior and the vagaries of sperm supplies. On the one hand, sperm are manufactured at a rate that could accommodate two sexual ejaculations a day. But if a man doesn't have that much sex, or at least daily sex, his body finds a way to get rid of the aging goods. Masturbation plays a role in making men's sperm supplies fresh, and ready for the next possible partner. On the other hand, too frequent sex, say more than once a day, puts an uncommon strain on the system. There's no time to replenish sperm supplies. And during that downtime, a woman might wander. So a man must guard his potentially wandering mate by increasing his chances of conception the next time they're together by ejecting a huge number of sperm. It's as if his testicles unconsciously "know" that a woman who has been out of sight might potentially copulate with another man. The Baker and Bellis studies point out that it's not the behavior or reproductive state of a woman that drives how much sperm a male contributes to her reproductive tract, but the lurking possibility of cuckoldry and dealing with his own dry spells.

Do Men Like Sex Better Than Women?

Sex researchers tracking attitudes about sex in Western society over the past several decades have noticed several trends. Men have more sex partners over a lifetime than women. More men than women approve of casual sex. Men have more extramarital affairs than women, although the women are catching up. When having sex, more men than women report having regular orgasms.[70] Taken together, these data seem to support the notion that men are more sexually experienced than women and they seek out sex more often. It seems reasonable to assume that men must naturally, that is biologically, like sex more than women. Evolutionary theory also supports this double standard. Men "should" be selected to spread their genes around, which of course means mating with fertile women. But since women don't go into estrus and it's impossible to tell who's fertile and who isn't, this means men "should" mate with as many different women as they can, as often as they can. Since women's primary reproductive aim is to conceive and have babies, there's no reason they should want sex very often, except to conceive. And so it's natural that men should be more sexually motivated than women—the sexes have been driven by natural selection to be different. Or are they?

DRIVEN BY MATING STRATEGIES

The predicted difference in mating strategies between men and women is based on how male and female animals, especially mammals, are able to pass on genes. At a primary level, the description of males sowing seeds and females protecting a limited number of eggs is correct. The scenario becomes much more complex when organisms have all sorts of options and constraints on top of those primary goals. And this is obviously true of humans. As a result, biologists and anthropologists have lots of room to disagree about how much men and women adhere to the traditional strategies of male promiscuity and female selectivity.

On one side of this controversy are biologists and social scientists who claim that men demonstrate their greater sexual interest at every turn. Anthropologist Donald Symons and psychologist David Buss maintain that men everywhere show that they want sexual variety more than women do.[71] Their data are convincing. Men do start having sex at an earlier age than women in all societies, men have a higher total number of partners over a lifetime, and societies worldwide are more lenient toward male promiscuity than female promiscuity. Men utilize pornography more than women. When asked, men also express a desire for a greater number of different partners than women do, although many men never have the opportunities or motivation to actually pursue their sexual desires. College-aged men routinely say they would have a one-night stand with an anonymous person, while women say that faced with the same opportunity, they would be reluctant to have sex with a virtual stranger.[72] Men admit they like sex for its own sake, while women often express a need for emotional intimacy as part of sex.[73] In other words, men generally like sex, and express an interest in sex with as many women as possible, while women are more selective and less sexually motivated.

But a different take on this information could produce a very different interpretation—that the differences are not "natural," in that men are genetically encoded to have more sex than women, but simply a product of sexual repression by males, a policy that developed to control paternity; without repression women would presumably have sex as often and with as many partners as men.[74]

Last summer, I attended a talk at a conference on human sexual behavior. The speaker was explaining a study conducted on a college campus in which attractive men and women were assigned to walk up to unfamiliar subjects and ask them to either have sex, go to their apartment, or meet them for a date later. As everyone expected, female subjects were less likely than male subjects to agree to casual sex. But what struck me was the smug smiles and laughs of the men in the audience—this confirmed for them that men are more interested in sex with strangers, more willing to have sex for sex's sake, than women. I pointed out that this study had a major flaw, and the speaker's interpretation was a little off. First, American women are taught from the day they're born never to go anywhere with a strange man. Safety is the issue here, and personal harm includes the possibility of rape, murder, and injury. This isn't a problem for men, so of course they might be interested in free sex when it's offered. For women, sex with a stranger is fraught with the danger of violence, and this is always added into the equation, at least in this country. Second, the speaker had missed the most surprising finding, the one that should have jumped off the pages. Some of these women would be happy to go to the guy's apartment, and a huge number said "yes" to a date. This suggested to me that many women, even in the face of possible danger, will go off with a stranger if he's sexually attractive to them. From my perspective, the study didn't reinforce a biological difference between men and women and their interest in sex with many partners; it said that women will be just as interested and might even place themselves in jeopardy for the possibility of a date and perhaps sex. I felt the view of the speaker, and the researchers conducting the study, was extremely culturally biased. In a less violent society, where women need not fear strangers, an entirely different result might occur.

The point is, the opportunities of sex with a multitude of others are constrained when you can't even talk to unfamiliar males out of fear. Humans have a long history of male domination. This probably stems from the male's need to sequester women away from other males to assure his paternity, especially when human infants require so much paternal investment. One cultural strategy to keep women away from other men has been to "castrate" women in the social and psychological sense, and turn them off to sex. Women might be interested in sexual variety, but this kind of behavior is made socially unacceptable in a patriarchal society. Women have been taught to be silent or uncommunicative about their sexual desires. Women are told again and again not to talk to male strangers. Women alone have their reputations ruined when the word gets out they've had sex with many partners. And the sexual interests of women have been virtually ignored in pornography. Perhaps by telling women over and over they don't like sex and restricting their sexuality, men have been able to culturally control what they can't restrain biologically—female sexual assertiveness. We can't blame men for this, because they have been acting in their reproductive interests. Even if women do have a natural desire for sex, or sex with many partners, perhaps their voices are not heard in this culture.

The major support for evidence of a cultural heavy hand on female sexuality is the ever narrowing gap between men and women in what women say about sex as they become more "liberated." More women now have extramarital affairs than when Kinsey and others were first asking questions about human sexuality in the 1950s. Girls are having sex at a younger age, and the number of different partners has increased dramatically for women who grew up during the sexual revolution.[75]

Also, when given a comfortable atmosphere in which to answer questions about sex, women freely talk about sex, the importance of a good-sized penis, their interest in sexual experience with a variety of men, and the need for orgasm.[76] One study on sexual desire showed that although the female subjects said love was important to sex, almost half of the women had had sex without emotional involvement, and 81 percent of the women (and 98 percent of the men) said they sometimes "needed" sex.[77] And in other cultures where female sexuality is acknowledged and talked about, women are expected to enjoy sex as much as men.[78] In other words, men and women may be less sexually dichotomized than some evolutionary biologists believe.

The only thing that really counts, from an evolutionary view, is patterns of behavior that are selected over time, and this only occurs when genes are passed on through fertile matings. The question is not how much sex each gender wants, but how much they actually get. And the sexes may be more alike in action than any of our fantasies might imagine.

Our gender, our age, our society, and who we are mold our sexual motivation. There probably are differences between how men and women enjoy sex, but as yet we really don't know how much of that difference is innate, or biological, and how much is culturally imposed. And we never will. In the end, men and women do eventually meet up and have sex, and they don't do this with just anybody. We neither mate at random, nor with only one person. We do, in fact, make choices for particular mates with whom to share our bodies, as we shall see in the next chapter.

Mate Choice

Ask a group of people, as I did, why they were first attracted to their mate and you might get the following answers: "I wanted to run my fingers through his curly hair"; "her smile"; "his blush"; "she wore cowboy boots and offered me a handful of M&M's"; "he made me laugh"; "he was from California"; "her enthusiasm about her work"; "I got a good recommendation"; "her sense of humor"; "his confidence, sense of humor, and red hair." These answers touch on just about every quality imaginable: looks, personality, geography, recommendations. I wasn't particularly surprised by the variety of the responses, or the fact that most of them are actually pretty useless when evaluating someone as a potential lifetime mate. After all, one of the more mysterious features of our mating system is what each of us is attracted to the first time we lay eyes on each other.

Nonetheless we *are* attracted to some people over others, drawn like a magnet in one direction rather than another.

Beyond first attraction are countless other strands that weave two people together into a mutual attraction and sometimes on to marriage and children. And so I asked the same people above why they stayed together. The answers were just as varied: "we're both silly"; "because we have the same level of sarcasm"; "we're good friends"; "we have exactly the same attitude toward other people, and ah, he's sexy"; "because I still like her smile"; "the tenderness we have for each other"; "our sense of humor"; "because of our dog." Most couples talked about their commonalities, an outlook they share, or similar personality traits. These answers point to, in most cases, a deeper connection between them of personality style and friendship.

From an evolutionary point of view, whom we decide to share our bodies, and our lives, with is perhaps the most important decision we might make. As a sexually reproducing species, we have no choice but to entwine our genes with someone else's. Sex with another person is how we do this, and interpersonal relationships are the context of that sex. Sure, there are quick matings that occur without much thinking or few consequences, but even these flings are not random. There are also matings that are done against the will of one party, and they also have reproductive consequences. But in the majority of cases, we freely choose our partners, and presumably these choices are based on *something*.

Theories of Choice

Being sexually aware of someone else is a complex affair. Upon introduction to new people, we take in their facial features, their body movements, what they say, and perhaps how they move. This is an instantaneous data-gathering mechanism conducted by the brain but orchestrated by the senses. We use our eyes, ears, and nose to pick up information about new people. Many times, judgments are instantaneous, and harsh. They can also be open-ended, positive, or neutral. But judge we do, and we do it unconsciously most of the time. While some might think this kind of evaluation is unfair, it's also part of what makes us humans, or animals. We need to evaluate and categorize all other humans because we live complex social lives. We are usually cautious about strangers; we need to know if someone is a foe or a friend. And if the stranger happens to be a member of the sex we're attracted to, we wiggle our personal antennae more vigorously.

There are at least half a dozen different academic theories on how we choose mates, falling into two general approaches. One group of theories points to biological origins; the other suggests personal psychology and the dictates of society are the rules by which we find our mates. These two kinds of theories address the same questions: Are our mate choices a product of selective pressures that have evolved over generations, or are they the result of more whimsical cultural and psychological options? Both views merit consideration.

THE EVOLUTION OF MATING

A group of researchers called evolutionary psychologists believes that the way we choose our mates is determined by desires hardwired into our brains.[1] Their approach is based on sound theoretical reasoning that goes like this: Making babies with someone is a major component of reproductive success. Surely the act of choosing a mate wouldn't be left up to chance. It seems reasonable to suggest that natural selection would have moved very quickly to make sure we chose the best possible mates, the best possible father or mother for our genes. This isn't necessarily a conscious decision. We can't, for example, really tell who might have "good genes" or predict parenting skills. We only have proximate clues like the way someone looks and acts to make a decision about our future. Therefore, we must have evolved mechanisms to detect the best partners from a group of possibilities, and the ability to choose "the best" mates.[2] If this is true, it would mean there must be universal, species-specific traits that identify what the best mate is to everyone. We see this behavior as falling in love, or being attracted to someone for no logical reason. But there is a reason, a reason our primitive brain compels us to choose certain mates, have sex, and pass on genes. In other words, our conscious psychology about choosing mates must have a genetic, that is heritable, basis that evolved over millions of years because it makes evolutionary sense.

These theorists then ask what those traits might be that would prove so important to mating and having children. The first assumption is that the two sexes must go about passing on genes in different ways. Males have many sperm to spread around and females have a strictly limited number of well-nurtured eggs and must invest heavily in their infants. It therefore follows that the sexes will also have different criteria for choosing mates. These differences in required parental investment dictate how each sex goes about finding mates and making their reproductive plans.[3] And because they have such different criteria for mating and parenting, it's a given that the sexes will forever be in conflict.[4]

According to this theory, men, with their low-cost, abundant sperm, "should" be interested in spreading their genes around. They should want sexual variety, a high number of partners, and be ready to copulate at the drop of a hat. But these men are constrained from such a "perfect" mating strategy by what women do and want. Women have to think about those intensely needy babies, and if men want their genes to make it to adulthood, they, too, have to be concerned about the nurturing of offspring. In that case, men will also choose the best females for reproductive purposes when they decide to choose a partner for the long term. Their mate would ideally be a fertile, healthy woman who shows no signs of having sex with other men. If he's going to invest his reproductive future in her, he wants to be sure any children that result from their mating are his. A woman, on the other hand, has an entirely different agenda. She wants a man who will provide for her and her children. She needs his money, his resources, and his protection to care for the children she's going to raise. The ideal man would then be someone of high socioeconomic status, someone with a lot of resources, or a rosy future.[5]

I present this theory not because I agree with it, or disagree with it, for that matter. I describe it here because this theory of men and women supposedly wanting very different things from their mates frames much of the recent research that's been done on human mate choice. These scientists believe that our individual psychology has been molded by evolutionary pressures as much as our bodies. According to this view, what we desire in a mate is a product of what worked best for our ancestors. Such theories expect men, like other male animals, to maximize their reproductive success by desiring lots of sex with many different women, and investing in offspring only when they have to. Under this theory, scientists expect women to be choosy about their mates, stingy with their sexuality, and very careful when faced with the possibility of sharing their genetic material. When women do finally settle on a partner, they expect he will be someone willing to commit to the long term and have sufficient resources or money; women should want older, established, higher-status men for mates. So when men say they philander, or when women express approval of rich men, say the evolutionary psychologists, their attitudes are only the result of our modern brains responding to ancient agendas.[6]

MY CULTURE MADE ME DO IT

Other social scientists oppose this evolutionary framework. They feel that personal psychology mixed with social pressure guides our mate choices.

Sigmund Freud believed that each of us seeks to fill a psychological void with our mate—we focus on the opposite-sexed parent and then look for someone just like Mom or Dad.[7] Although this view has dominated Western psychology for many years, no one really knows how applicable it is to Western culture in particular, or to humans universally. Another possibility is that we use our mates to fill a different kind of void —to find qualities we personally don't have.[8] In this view, opposites attract, pulled together like magnets. This exchange of opposites, a Marxist might also suggest, is economic and mutually advantageous.[9] Some believe in the reverse side of this—that we have a reasonable evaluation of ourselves, the way we look, and our own value system, and seek someone who is similar to us to avoid conflict.[10] This notion, "the matching hypothesis," has gotten the most attention and real-life testing by social psychologists, and it looks like similarity in personality, education, intelligence, and so on is indeed an important part of whom we choose (see below).

Other opponents of the evolutionary approach to mate choice include social anthropologists and feminists who believe society, not genes, governs whom we choose as mates.[11] For example, most couples are pushed and pulled by what society, parents, or peers think is the "right" step. In this case, our notions of the best mate are sculpted by the culture around us. Women are especially vulnerable to the dictates of family and society. In almost all societies around the world, women have little economic or political power. And so women often want husbands with high socioeconomic status because women usually can't get money and power themselves. A cultural anthropology colleague of mine, however, pointed out, "The sentence 'Women want men who have power and money' works just as well if you erase the words 'men who have.'" Her point is well taken. At some time in our history, men gained all the power and resources. We can't assume that male domination is a hardwired feature of human societies. After all, not all female primates are dominated by males.[12] Instead, male power arises when resources can be sequestered. When males must compete with each other to gain resources, they grow larger, stronger, and possibly more aggressive. Females, on the other hand, remain the optimal size for gestating and nursing infants and can't energetically afford to expend energy on large body size.[13] As a result, in some species males are physically much larger, and therefore more imposing, than females. If they want to, they can physically dominate. Females then become one of the resources that males sequester, and in the case of human females, they lose the ability to attain power themselves.

Our nearest relatives, the chimpanzees and bonobos, illustrate two very different scenarios for the ancient mating system that we and those apes might have shared. Chimpanzee males dominate females, and are sometimes brutal toward them. Females have their own sorts of power in mating, but in general, primatologists would agree that a male chimp is most often dominant over a female chimp.[14] Bonobos, on the other hand, lead a different sort of life. No other bonobo tells a female bonobo where to go or what to do. Bonobo society is egalitarian in food distribution, sex, and status.[15] It's impossible to say which kind of system our ancestors had 14 million years ago when we shared a common ancestor with chimpanzees and bonobos. But it's reasonable to suggest that females were less overpowered than they are today.[16] If so, there's no telling when our society became male-dominated. It may have occurred when humans stopped forming into hunter-gatherer bands, and men, who were a bit larger, began to horde resources—land and women—by fighting off other men. Women became chattel because they had no choice. In any case, men hold the majority of economic resources throughout the world today, and no one would argue with that.

Opponents of the evolutionary approach counter the evolutionary psychologists not on the issue of male domination today, but on the source and effect of that domination. The social construct view suggests that women want men of high income and status, but not because women are genetically predisposed to do so. Culture, not biology, forces women to mate with men who can give them what they want and need because they can't get it any other way.[17]

The evolutionary approach is also criticized by those who conduct long-term studies of mating and marriage.[18] Such scientists maintain that the evolutionary approach is superficial in that studies try to decipher the exact physical or personality traits that initiate a mate choice while ignoring the long-term consequences of marriage and family formation. Relationships, these researchers contend, are not one-time decisions based on discrete traits. Marriage, if not mating, is a process, not an act. And that process is a long series of negotiations. They also point out that many traits change over time; marriages that might start out based on one trait —looks, for example—always evolve into something else over time. In this view, shallow characteristics, like the attraction of a smile or an exotic background, are only clues to the person underneath.

It's possible to reconcile the evolutionary and social construct approaches to mate choice.[19] We, like all animals, have reasons for mating with one person over another. Part of our behavior can be explained by a compelling urge to pass on genes in the most efficient way. Our instinctual urges for this or that person might be directed by our genes, helping us to focus on proximate cues that accurately or inaccurately point us to the "right" mate. But of course, we don't always act in a way that

WHAT'S LOVE GOT TO DO WITH IT?

benefits the passing on of genes. We make mistakes in mate choices, evolutionarily unsound decisions, and have sex at times with people who are inappropriate on every level. And so we are often betraying our own inner voices, voices that get it wrong half the time anyway. At the same time, our environment, too, strongly influences whom we choose as mates. Acculturation and socialization may explain those features of human mating that vary across cultures, and society sometimes directs us in ways that are not necessarily beneficial for reproductive success per se. Given our genetic urges and the cultural pressures put upon us, our mate choices and reproductive decisions don't always fit into a nice neat package that can be quantified and analyzed by scientists.

What Are We Allowed?

A student once told me that there are fifty thousand appropriate partners in the world for each of us. Where she got this number I'll never know, but it does make some sense. Given the number of people in the world, I can easily imagine there are at least fifty thousand that I could fall in love with. Unfortunately, I will never run into most of them. I am stuck, as all of us are, with a small social circle, what scientists refer to as our culturally and geographically defined mating pool. Within that pool, every society has rules governing or pushing people to mate with certain others.[20] One such rule prohibits incest. Other rules of conduct apply to the race, class, or religion of possible mate choices. In this way, the larger pool of those available is whittled down to a much smaller available group; those fifty thousand possibilities quickly diminish to five thousand or fifty, or maybe just five.

HOW WE MATE

People have sexual partners both in and out of the context of marriage. These are, of course, human constructs. While dividing mating into two distinct categories is somewhat artificial, it is important in the human context because of the risk of pregnancy and the consequences of infant care. Marriage is the legally sanctioned, publicly celebrated and acknowledged form of human mating.[21] Marriage also implies, and encourages, sexuality. In all cultures, marriage means sex which means children. In other words, marriage is the most common form of human mating and the usual route for parental investment. Even when there are no children, marriage vows represent a commitment to, among other things, sexual access. Of course, sex and children are not the only functions of marriage. It is also an economic unit, and at times a political unit as well.

The other form of human mating—sex outside the bounds of marriage—takes place before, during, and after marriage. Cultures differ in the amount of premarriage sex they allow. Anthropologist Suzanne Frayser found that people in the far Pacific Islands are most permissive toward adolescent sex, most African and Eurasian societies are reasonably tolerant, and only certain societies that lie geographically around the Mediterranean Sea are highly restrictive toward premarital sex.[22] And in general, there isn't much of a double standard between the sexes in how societies see sex before marriage; it's usually acceptable for males as well as females. In our own culture, it's now common, and even expected, that both young men and women will not be virgins when they marry.

Rules regarding casual sex change, however, once the bonds of matrimony come into effect. But still, some men in at least 80 percent and some women in 73 percent of all cultures worldwide have admitted to adultery.[23] Chances are, these reported numbers are actually much lower than the real incidence. When Alfred Kinsey and his colleagues set about asking Americans in the 1950s about their sex lives, people most often refused to be in the survey when they knew there would be questions about extramarital sex. And this question, the one on infidelity, was the one most often not answered by the subjects.[24] Men reportedly have extramarital affairs more often than women, but in all societies, some married women are involved.[25] And as I mentioned previously, female philandering is of special interest because there's also strong disapproval against women in almost all cultures.[26] Given that, it's rather amazing that any women risk scorn and sometimes severe punishment to have sex outside marriage; it certainly speaks to the strength of the female sex drive.

Extramarital affairs are risky in terms of reproductive success. In several studies using blood samples to determine paternity, genetic markers showed that at least 10 percent of the infants had fathers other than the men designated on the birth certificate.[27] And a recent survey conducted in Britain showed that women not using birth control often had sex outside of their committed relationships right around the time of ovulation, putting themselves at risk for conception.[28]

What isn't often known is the quality or character of the people chosen as extra partners. We can dissect and evaluate marriage because it is socially acceptable and usually documented, but as yet, no one knows much about the particular partners chosen on the road to sexual deceit.

NOT TOO CLOSE TO HOME

There's no such thing as random mating. There's also no such thing as free choice. Society and evolution conspire to form our mate decisions, and the first law of mating is—don't choose too close to home. Across every human culture, in every geographic location, no matter the social structure, every human society has a taboo against incest.[29] Incest avoidance is also the rule in animal groups. In other primates, for example, there's always one sex or the other that leaves the group at sexual maturity.[30] Female chimpanzees move away from their home group at about the age of fifteen and try to immigrate into a neighboring troop;[31] macaque males, as they mature, usually wander off and find a new home;[32] howler monkey males and females both leave home and seek new troops.[33] This emigration means that, by definition, most close relatives aren't around to serve as available partners. Even if they do remain in proximity, animals seem to ignore the advances of close relatives, or somehow prevent matings.

For humans, the rules against incest are sometimes unspoken, sometimes made explicit by oral tradition or written laws. In both cases, those who violate these rules are punished. At the same time, these rules vary across cultures; one culture's incest avoidance is another culture's mating opportunity. In Western culture, for example, incest includes sexual relations extending out to first, and sometimes second, cousins. In other cultures, incest draws a wider circle—no mating is allowed within clans or communities. These rules also change over time. Only one hundred or so years ago, Charles Darwin married his first cousin Emma Wedgewood, and this was considered perfectly correct. Today this kind of marriage would be a scandal, splashed across the cover of a tabloid magazine. And sometimes, when it suits the purposes of the social group, the rules of incest are broken. Many an empire has been fortified by marrying off close relatives and linking two kingdoms. Obviously, these kinds of marriages have more to do with power brokering than mate choice, but they do demonstrate a certain social flexibility in the incest avoidance system.

Some have suggested the human incest taboo is different from incest avoidance found in other animals because humans "think" before they mate, and therefore, the human reaction to close relatives must have a psychological basis. Freud believed the taboo arose from a deep conflict within each of us to mate with our parents. According to Freud, long ago our primitive ancestors did mate with their close relatives. In particular, young men killed their fathers and had sex with their mothers. But this event produced such guilt that the incest taboo was instituted, and handed down to more sophisticated societies. Freud felt aboriginal societies were more "primitive" and so their feelings and impulses were close to the surface. Their taboos against certain mates were strongly stated and highly complex because these groups were closer to their baser, destructive instincts than more "civilized" societies, which didn't need such stringent rules.[34] As human groups became more civilized, Freud reasoned, they were more emotionally in control, and the taboo needed only to be enforced for cousins.

Obviously, Freud's scheme is culturally offensive. It has also never been substantiated for any group of people, aboriginal or otherwise. Bronislaw Malinowski, the first anthropologist to evaluate the universality of the incest taboo, showed that the incest taboo follows the lines of kinship and authority.[35] Wherever authority is threatened, or family ties are in jeopardy, there is a strong, strictly enforced, incest taboo. According to Malinowski, sexual relationships are forbidden where they might cause conflict. Claude Lévi-Strauss proposed a much more Machiavellian explanation for the incest taboo.[36] He pointed out that by definition, incest rules force group members to make exchanges, and therefore alliances, with outsiders. The incest taboo, according to Lévi-Strauss, is a significant social strategy.

Biologists take a more pragmatic stance concerning the incest taboo. They assume that incest avoidance evolved in all animals as part of mating systems because those who mate incestuously run the risk of combining bad genes. Each of us carries a certain genetic load of aberrant genes. These genes cause no harm because they're in the recessive state, and masked by a matching normal gene. But when that aberrant gene is combined with another bad copy, as it could be if a conception occurred with a close relative who was also carrying the same gene, the two aberrant genes would express the dominant devastating trait or disease. The long-term lethal effects of inbreeding, when closely related individuals mate and conceive offspring, has been selected against over time in all sexually reproducing species. And so, monkey mothers instinctively avoid mating with their sons, sister chimpanzees aren't interested in their brothers, and close human relatives are restricted from marrying. The incest taboo, in this view, is natural selection's way of filtering the gene pool. We, like all animals, instinctively follow a simple rule—don't mate with those we grew up with. This rule, often called the Westermark Hypothesis after the first person who recognized that familiarity breeds disinterest in mates, is best demonstrated by the studies of Israeli kibbutzim, in which there are few marriages among nonrelatives who grow up in close quarters.[37] There's no reason these young adults couldn't fall in love. They just don't. Nature presumably has selected for a genetic complement that gets turned off by the sight of someone we know all too well. In this way, we "naturally" avoid combining our genetic material with a set of genes that might carry the same genetic complement.

There's no doubt that all these explanations have some validity in explaining mating in human groups.[38] The more social explanations of resource exchange and group cooperation are certainly valid. But underneath the layers of cultural explanations for human mating is the fact that all other animals share with us this drive to mate *outside* a close circle of relatives, a fact that automatically avoids genetic inbreeding.

KEEP WITH THE PROGRAM

Societies typically have a consensus view about who should marry whom. Think of the phrase "mixed marriage," which inclusively refers to a marriage of people of different races, different religions, or even different social classes. No one knows exactly who instituted societal rules, but they often determine whom one is exposed to, attracted to, or even allowed to marry. In a cross-cultural analysis of mating rules, anthropologist Suzanne Frayser found that kinship is the major criterion by which all societies make their marriage rules.[39] That is, beyond the incest taboo against certain mates, there are other directions that often must be followed. In some cases, one must marry another of the same clan, matriline, patriline, or, in some societies, quite the opposite—these are just the people you're not allowed to marry. Interestingly enough, those we choose for extramarital affairs tend to come from the same pool of possibilities.[40] Frayser notes that all other criteria, such as age, locality, or tribal affiliation, pale next to the rules of kinship. Looking at sixty-two societies, Frayser found that the number of societies in which a person was required to mate outside the community, or inside the community, or where it really didn't matter was about equal. In other words, there's no universal rule, or even tendency, to restrict possible mates in any particular direction. Certainly we see a type of nudging toward within-community marriage in American society when parents encourage children to marry "their own kind."

Society also influences interpersonal relationships when a type of marriage is either favored or forbidden. In America, for example, most states, except Utah, ban marriage to more than one partner at a time. But in about 84 percent of cultures worldwide, polygyny is allowed or favored.[41] People in Western culture equate polygyny with sexual excess, and are shocked. In fact, polygyny is really an economic system and one that's available only to men of high status and income. In some societies, a man who is rich enough can take on a second or even a third wife. Additional wives will give him more children, prestige, and helping hands. This man is by no means promiscuous, however—he's just having sex with two women on a regular basis instead of one. In fact, this type of marriage system might be viewed as an economic burden to males; the polygynous husband has more responsibilities with a larger household. In any case, the polygynous marriage, although often favored or accepted, is actually rare. The vast majority of marriages in the world involve one man and one woman.[42]

This doesn't mean that we are "naturally" sexually monogamous, however—or that most people stay with one person all their lives.[43] It just means that the usual marriage system for humans, the system in which most children are born, is the pair-bond. A polygynous household with one father and a few mothers is much more rare.

ARRANGED MARRIAGE

The heavy hand of society and family is most clearly demonstrated in societies in which marriages are arranged. In looking at information on marriage patterns in 133 cultures worldwide, marriage by arrangement, that is, when parties other than the potential bride or groom have a say in who should marry whom, occurs in 106 (80 percent) of the cultures I surveyed.[44] The ethnographic reports of marriage systems, however, showed me that arranged marriage was the sole way families are started in only 23 cultures (17 percent). Suzanne Frayser found the same result; when marriages are arranged, three quarters of the time grooms and brides are always consulted, and they are not forced into marriage against their will.[45] It seems even when groups plan a marriage for others, they make allowances for other kinds of arrangements. And even more interesting, when men and women are allowed a say in such a marriage, both parties, the prospective bride and groom, typically have a say in the plan.[46] The image of a girl married off to a stranger that she doesn't want is a fallacy. Even in those few societies that decide on husbands for their daughters while they are still infants or children, the very same adult daughter is free to break that engagement in 25 percent of these groups.

Our picture of arranged marriages ought to be adjusted. There's no reason to assume "arranged" means forced. In only seven societies in the Mediterranean area and a few in China, the same societies that are so restrictive about sex, are women not consulted at all about their future.[47] In most cases, an arranged marriage means the family is often only formalizing what two people have already decided, and women and men have all sorts of ways to influence choices made by others. Even when the partners are strangers, we can't assume this choice imposed on them by others is unwanted or bad.[48] There's always room for expectation and fantasy; at least an arranged marriage isn't based on such transitory emotions as love and sexual attraction, as most marriages are in Western cultures. Given that marriage is really a system designed for having children, rather than a union for sexual fulfillment, perhaps an arrangement made by others is just as good, or even better. And let's not forget that marriage doesn't necessarily mean that the woman, or the man, will have sex thereafter only with the partner placed in his or her company.

In our own culture, marriages aren't usually arranged in the formal sense, but nonetheless they are constrained by the dictates of society. Most of us operate within a circle of friends and acquaintances of the same race, education, religion, and socioeconomic class. While marriages in America certainly aren't formally arranged, there *is* a kind of setup.

What Does Everybody Want?

What's your "type"? friends may ask you. Is he tall or short, dark or light, sexy, emotional, trustworthy, risky? Is she pretty, young, older, smart, vulnerable, strong, sexy, impulsive? Left to our own devices, every person has some sense of what they might look for in a mate. I could list a hundred adjectives and let you choose which words best apply to your ideal partner. But there is no need to, because evolutionary psychologists have spent literally hundreds of hours trying to figure out what we all want. These studies are based on a simple idea—since mating is so important to reproductive success, natural selection must have given us all some mechanism for evaluating others as possible mates.[49] In this scenario, these unconscious, or even conscious, antennae have been molded by evolution to spot what is best for each person's reproductive future. And that future involves figuring out the reproductive value of the other person. We see our attraction as romance, lust, an attraction to a specific individual's chemistry; evolutionary psychologists see it as evolved unconscious strategies that help lead us in the direction of the "right" person.

WHAT WE ALL "SHOULD" WANT

The evolutionists argue that each sex should be expected to look for partners of high reproductive value. By the value of the partner they mean what he or she has to offer to help advance our genes. Men, for example, should want fertile women. And women should be attracted to men who can help them bring up any children they produce. This is the bottom-line version of the theory advanced by evolutionary psychologists, based, as I've said before, on how much each sex invests in gametes and offspring, with men investing little and women investing a great deal.[50] A more elaborate version of this paradigm suggests that women should be concerned with a man's wealth, his status, and his earning potential, because these are the resources that she wants for her children. And men should be interested in finding a young woman who is more likely to have a long reproductive life before her. But how to tell if one woman is more fertile than another? The face might be the obvious place to look for signs of health, youth, and fertility. And so researchers have suggested that there is a universal standard for female beauty that signals to males that a woman is young, healthy, and fertile.[51]

Something inside me, however, rebels against this hypothetical difference in the ways men and women are supposed to evaluate each other. It's not that I disagree, exactly, but rather that I don't believe that the sexes are so polarized in what they are looking for in a mate. Women who want to improve their reproductive success should also want a man with a long life ahead of him, a young, healthy, well-complected, and handsome man. Also, women can get help in raising children from others besides a male mate. A woman can also possibly make it on her own, rely on relatives, or other members of her community. And to say that men should only want young women actually runs contrary to an evolutionary view—I would think the "best" reproductive partner for a man would be a woman with proven fertility, a woman who has already borne one child. And finally, prettiness has little to do with health or fertility—a woman who looks like Cindy Crawford could be infertile and a woman who is plain or heavyset might well be exceptionally fertile. In fact, many of our modern ideas about beauty, such as thinness for women, are contrary indicators of fertility. If our standards of beauty were biologically linked to fertility, there would be no change over generations in what is considered attractive, nor would there be differences across cultures.

ATTRACTIVENESS IS IN THE EYE OF THE BEHOLDER

Attraction between two people usually starts with looks; physical attractiveness, more often than any other character, brings people together of their not-so-free will. The problem is figuring out exactly what "attractive" means to one person or another. In general, attractive people are considered more desirable than plain or ugly people. Anthropologists maintain culture plays a heavy role in what each group defines as beautiful.[52] For example, fatness is considered not particularly attractive in our modern society, but it is considered a sign of wealth and prosperity in other cultures. Our standard-bearers of modern-day American beauty, movie stars and fashion models, are often painfully thin. But groups such as the Chuckee, Hidatsa, and Thonga believe beautiful women should be powerfully built.[53] And think about the myriad ways people over the world alter their bodies to emphasize what they see as beauty. Tattoos, lip plugs, stretched labia, penises covered in gourds—people do all sorts of self-manipulative things to their bodies to increase their attractability. Also, these tastes change over time. What was beautiful, fashionable, and attractive twenty years ago may look rather silly today. Take large sideburns on men. They've come and gone in fashion many times over the past hundred years, so frequently that men who have them ought to keep a razor handy. Anthropologists point to the evolution of attractiveness, its different criteria among cultures, and the fact that outside influences, such as the media or fashion industry, can manipulate what people think is attractive to support their view that beauty is culturally molded and has nothing much to do with anything evolutionary.[54]

But there are some traits that all societies consider alluring. Cleanliness and health care are usually connected with attractiveness. Signs of illness, such as bad skin or a cough, usually repel potential mates.[55] Good social skills always make someone seem more attractive.[56] Cross-culturally, men seem to be more attractive when they are perceived as skillful at their tasks and powerful in status, while female attractiveness is more centered on physical properties.[57] Attractiveness in general is, however, not just a question of a pretty face, or a nice body, but everything about the person in question. In that sense, attractiveness does not always mean being handsome or pretty. It really means being sexually attractive, which is made up of many physical and personality features.

FACING IT

What is beauty? While no one denies a cultural component to beauty, there's also a suspicion on the part of scientists that somewhere deep in our genes we all have a similar standard for what is "pretty." To examine this possibility more quantitatively, researchers used a scanner to digitize the faces of ninety-six men and women into a computer. They then meshed these faces and randomly made three sets of thirty-two faces each.[58] These new composite faces included a mix of Caucasian, Asian, and Hispanic features, all with the same neutral expression. The computer then generated a series of faces made up of two faces mixed together, then four, eight, sixteen, and all thirty-two. When the researchers showed these faces to a large audience and asked for ratings of attractiveness, the composite face made up of the most variations, the entire thirty-two, was picked as the most attractive. In other words, everyone, when given the option, likes an average, composite human face.

But sometimes there's the possibility of a better than average composite human face. Other researchers scanned into the computer photographs of sixty-eight female faces to give the computer a population to work with.[59] These faces were then broken down into parts—hair, lips, noses, mouths, and chins—resulting in 17 billion possible combinations. And then they let people loose on the computer and allowed them to play God. Both men and women, in attempting to put together an ideal human face, settled on roughly the same pretty female face, although not everyone was confident of their handiwork. Unlike the previous composite, this face was only "average" in the eyes. The more "ideal" computerized female face these men and women created had a more pronounced forehead, and a very short lower face. In particular, the area from nose to lip, and lower lip to chin, was extremely short compared to the average. In fact, the measurement of this "perfect" face from nose to chin best matched the faces of eleven-year-old girls. Also, the lips were rather full and the width of the mouth very short. In other words, the ideal female face to these American subjects was in essence a baby-doll face.

What about the male face? Based on the assumption that women should want mature, approachable, but high-status males, researchers took sixty black and white photographs of men and measured their various features. They then asked one hundred women undergraduates to look at various faces. Women selected an idealized man with large eyes, a small nose, prominent cheekbones, and a large chin. The women didn't think highly of certain features, such as a receding hairline or thick facial hair, that would signal maturity.[60] The ideal male to them had a mixture of youngish features that the researchers interpreted as indicative of an open, nonthreatening individual, and a certain raggedness.

More recently, researchers again tested the "averageness" hypothesis by making composites of various faces and mathematically rendering all the features of each face.[61] They then computerized a series of three female faces. The first face was an average composite based on all the faces included in the study. The next face was made up of an average with certain traits that were exaggerated based on those female faces that had been judged most attractive before they were computerized. If the most attractive fifteen women had full lips, for example, the average face got fuller lips to make her more attractive than average. This face was then exaggerated even further to create a third, supposedly hyperattractive face. When different subjects were shown the three faces, everyone, men and women alike, preferred the hyperattractive face the most. The results of these tests seem to indicate that given the chance, everybody picks beauty over averageness.

None of this seems particularly surprising. We might bicker about the fine details of someone's attractiveness, but everyone knows a pretty face when they see one. We also know that one reason this face is so beautiful is that it's very unusual in terms of the population as a whole. If every man looked like Mel Gibson, his handsomeness might lose some of its appeal and not have much meaning. But there's only one Mel Gibson. And so models are paid high fees simply because such agreed-upon perfection is scarce in our species. Ideal beauty, then, is something we all acknowledge, and perhaps desire, but few of us obtain such beauty in our partners. And even if we did, that beautiful face might turn out to be a hindrance, because everyone else would want our partner too. Most of us, if we're lucky, fall in love with something close to the average composite face.

BELOW THE NECK

It's an unfortunate fact that just about every woman in American thinks she's fat. The roots of this identity crisis are complex, but probably have something to do with how the media present women to themselves.[62] When asked, women always see themselves as fatter than they really are. And more important, their image of what they think men like is equally distorted. Men repeatedly find attractive women with much more bulk than women think is the ideal.[63] This might be because fatness or thinness really isn't the issue, for men at least. Curious about whether or not we judge people not just on the basis of their face, but on their body shape as a whole, scientists conducted a recent study that looked at the shape of females—not just their weight—and how men reacted to different female figures.[64] The study was based on the evolutionary thesis that a certain body shape is a sign of health and good reproductive success. Before puberty, body shape in girls and boys is rather similar, lean and straight without many curves. Adult men and women, however, have distinctively different body outlines. Women, under the influence of estrogen, deposit fat on the hips and thighs, and men, under the influence of testosterone, deposit fat at the abdominal region. The result is that most men tend to be broad-shouldered and narrow-hipped, with extra fat accumulating at the stomach region, and women tend to be small-waisted and wide-hipped. In all human cultures, men have high waist-to-hip ratios and women have low ones, and there is no statistical overlap in this number between men and women. More significant, these researchers claim, is that the distribution of body fat has reproductive consequences, at least for women. Women with a high waist-to-hip ratio, say over 0.80, usually entered puberty later, have lower fertility, and are susceptible to diseases such as diabetes, heart disease, stroke, and cancer of the ovaries, breasts, and endometrium. Pregnant women have an extremely high waist-to-hip ratio, way over 1.00, and while this might signal the possibility of fertility in the future, men might associate such a look with pregnancy, which would negate any chances for immediate conception. Women with high waist-to-hip ratios, in other words, "shouldn't" look particularly attractive to men interested in improving their reproductive success.

To test the notion that waist-to-hip ratios might be an indicator of health and fertility, 106 male subjects were given a series of line drawings of women of all sorts of weights and shapes. Some drawings were of very thin women with low waist-to-hip ratios while others were heftier women, also with low waist-to-hip ratios. Other outlines included thin women with thick waists and fat women with great curves. The men consistently chose the middle-weight figures with the lowest waist-to-hip ratio. This study is intriguing because it focuses not on facial beauty, which is so clearly molded by a cultural norm, but the shape of the body, which is more clearly tied to fertility. While a baby-doll face might constitute a male's idealized face, that eleven-year-old archetype is not the best for reproduction; for example, the infant mortality rate for young girls is extremely high. It makes better biological and evolutionary sense for men to be attracted to a fully mature female body, one that signals health and fertility, a shape that he can even see from the side or the back. As yet, however, no one knows what females may think the ideal male body type might be. Perhaps we may find there is some best fat distribution in males that signals to females high testosterone and good sperm counts or good parenting skills.

WHAT EVERYBODY SAYS THEY WANT

When asked, people everywhere tend to agree that kindness, compassion, and caring are among the most important traits a mate can have.[65] The data are also pretty clear about the differences between what men say they want and what women say they want. Men of all ages, from college age to older men—and men in almost every culture—say female attractiveness is extremely important to them when thinking about choosing a mate. That is, when presented with a list of possible traits to consider, men rate physical attractiveness higher than women do. They also most often want someone younger than they are. Women, on the other hand, say looks are less important to them; they want a man who has good financial prospects, as well as a good personality, someone who is emotionally involved and will offer a sense of commitment.[66] A man's status and finances are so important to women that a man's attractiveness improves when she thinks he has a good future, while a female's attractiveness is improved only if she is perceived to have a likable personality.[67] Further proof that women look for status and money in a mate comes from several studies that asked female medical students, women with good financial futures, how important financial stability was in a prospective mate. Even though these female pre-meds would surely have status and financial security of their own, these women said that their ideal mate would have even higher status and money.[68] According to these studies, women think their mates should be capable of bringing in more than they do, regardless of what they themselves make.[69]

And apparently, some women and men know how to advertise those qualities the opposite sex is looking for. In two analyses of singles advertisements placed in newspapers and magazines, women offer descriptions of their attractiveness while mentioning that they would like someone with a decent job. Men more often advertise that they are looking for an attractive younger woman, while offering financial stability and expressing a desire for a long-term commitment or marriage. Both sexes assure the opposite sex of their sincerity of character.[70]

These statements of what it is men and women desire make sense to evolutionary psychologists. At first glance, they confirm that each sex is seeking something of reproductive value in a mate. Men, who want to spread sperm around and mate with fertile women, should be interested in young, healthy women with long reproductive lives ahead of them. In evolutionary psychologists' view, when men universally say they want a young, pretty wife it supports their notion that men have been selected over generations to pay attention to proximate cues, that is easily observed cues, of health and fertility. And when a woman says she wants a man with status and money, she's only looking for those traits that will help her raise dependent infants.

These theorists also recognize that mating, or sex the act, occurs in two different situations. Sometimes sex is a short-term affair, and at other times, the sexual relationship lasts a lifetime. Accordingly men and women might alter their mate choices for these two different types of mating. When asked, men said they would readily lower their standards for beauty, and other qualities, for a chance at a one-night stand.[71] Women, on the other hand, tend to have similar standards for both short-term liaisons and long-term relationships. This shift in standards in men again makes evolutionary and reproductive sense because men should first and foremost be interested in spreading their sperm around. But women, who have more to lose if they get pregnant even by "mistake," need to be more choosy regardless of whether the sex is quick or part of a committed relationship.

And so, say these theorists, when people are presented with a list of what they might want from the opposite sex they line up like opposing teams on a playing field: Men say youth and attractiveness are the important issues in selecting mates, and women focus on status and wealth.[72] It's all a nice, neat package. Or is it?

WHAT'S WRONG WITH THIS PICTURE

Ask me what I want in a partner and I'll give you this list—an emotionally stable man about my age, successful and happy with his career, who has a decent income, is responsible, cute, comfortable with his body, and of course, has a good sense of humor. Who wouldn't want a partner exactly like that? The first problem is that in real life few humans fit any fictional list of the perfect mate. If they do, there's a good chance us regular folks won't be on *their* list. I call this the Tom Cruise/Julia Roberts phenomenon. Yes, most women would like to have a cute, rich, young Tom Cruise for a mate, and most men would love a cute, rich, young Julia Roberts for a wife or girlfriend. But this is more about what we'd like in our fantasy world rather than what we end up with in the real world. The first and foremost problem with asking people what's important to them in a mate is that wishes are not real life. Even anthropologist Donald Symons, who is committed to the idea that men are evolutionarily hardwired to desire young, pretty women, once said in a presentation, "Any fifty-year-old man who thinks he will ever get to be with a twenty-year-old woman is deluding himself."[73] And those women who think they will marry men of high status and income who will support them and help bring up their children are probably living in a fantasy world too. There just aren't all that many young, pretty, fertile women or high-status men available, and so almost everyone ends up with a mate who is quite different from the ideal. In the end, just because people "say" they want something in a mate does not necessarily mean their fantasies are echoes of some sort of genetic predisposition. Life just doesn't work that way. Natural selection operates by selecting for certain patterns in morphology and behavior. The evolutionary psychologists think that when people in all sorts of societies say they desire young women or high-status men, these pronouncements are evolved strategies. But a strategy isn't a strategy if it's rarely implemented; it's just an unfulfilled wish. And evolution doesn't recognize unfulfilled wishes. In other words, what people say they want is not important, it is what they get that is of real significance.

And perhaps all this talk of young, pretty women and older rich men is off the track anyway. In all studies of ideal mates, men and women say that the first, and most important, factors they consider in choosing a mate are love, mutual attraction, kindness, or other personality traits.[74] Men and women are in striking agreement about this—it's the qualities of a partner, not their looks or financial statement, that matter most. It's not until after the personality traits are rated that other traits come into play. So one way to look at the data on mate choice wish lists is that the sexes are more alike than different. While evolutionary psychologists admit that personality and mutual love are the most important qualities to everybody, they choose to emphasize the differences between the sexes rather than the similarities. In one extensive study of mate choices, psychologist David Buss and colleagues gathered data on a list of eighteen possible qualities of a mate. They interviewed 10,047 people in thirty-seven different cultures and asked them to rate those eighteen characteristics for importance in their ideal mate. The highest-rated characteristics almost universally for both sexes were, in order, love, dependable character, emotional stability, and pleasing disposition. Only when the list got to number five did the sexes diverge down their different paths. Also, how men and women rated the importance of possible mate characteristics varied more between cultures than within them.[75] That is, the wish lists of men and women from the same culture tended to look more alike than the wish lists of either women or men across several cultures. It's only in the few cultures where women are sexually sequestered or repressed that wide differences become apparent; virginity in particular is extremely important to men in these cultures, but not to women.

Another difficult issue is figuring out how much culture molds what these psychologists see as ingrained desires. Surely, men in our culture are bombarded with images of pretty, young women as the ideal female face and figure. We can't tell, however, if men respond to these images because their inner biology tells them to, or because they have been told since infancy by the media and their culture that this is the ideal. Women have also been taught since childhood that they will never make as much money or have the same status and income as men do; they learn the easiest strategy for attaining wealth or rank is to marry it. Women have also been taught that if they follow their natural sexual desires, they will be punished. It's much easier, then, to follow the socially dictated path.

THE PATH OF MATE CHOICE

There's no question that there's a certain dichotomy between what women and men want in a mate. There's also no question that the sexes are very similar—we all want someone to love who isn't a loser. The important question is the root of these desires. There is a deep biological motivation in each of us, whether we respond to it or not, to pass on genes. Because we are sexual reproducers, this means finding a mate, and because we have such dependent offspring, it also means long hours of parenting. It seems reasonable to suggest, as I have previously, that our biological layer is attuned to this, and it should be as true for women, who have those babies, as it is to men, who have to invest too.[76] Women need not manipulate men into staying with them, because natural selection has done this for them. If parental investment in general is so important for a human child, and it takes more than one adult to invest, that extra parent will most likely be the father. If he doesn't stay, the child's future, and therefore the man's reproductive success, is at risk. Although women could bring up children alone, or with the help of kin, this is a shaky strategy because kin have less to lose, biologically, than the actual parent. So if a man wants to pass on his genes, he has to stay and invest. This isn't manipulation by women, but a choice by men. It's no surprise, therefore, that people say the most important thing to them in a prospective mate is a personality they can get along with. If we look at "mateships" as long-term relationships in which children are nurtured, rather than as a quick pickup, people should look for a mate who is a good negotiator, someone they can work with.[77] An overriding need for cooperation and negotiation may have made the sexes much more alike than different in their mateship desires. Beyond that biological layer, culture has conspired to give men the upper hand in owning the resources, and so women must adapt.

Mating in the short term is very different, but it also has commonalty for both sexes. In the short term, when women have the opportunity and freedom, I maintain that women are just as interested in several partners, and frequent sex, as men. One study by Heidi Greiling of the University of Michigan showed that when men and women were asked about their reasons for having one-night stands, both sexes cited "sexual pleasure" as the major incentive.[78] And so here, more than anywhere, is where I differ from the evolutionary psychologists.[79] I think that the real conflict for humans is not between the sexes, but within the individual. For men, an inner voice says, have sex with as many partners as possible, and take a chance that you might impregnate someone, get lucky, and have your child brought up by someone else. Another voice says, stay with one woman and at least make sure some of your offspring survive. For women, the same voices call out. Stay with one man and make sure your children get all they need. But meanwhile, enjoy all the sexual possibilities that are out there; your culture may not approve, but do it anyway. And so for each of us, long-term and short-term relationships are always a compromise at many levels, no matter what your sex is.

What Does Everybody Really Get?

Whom do we actually make babies with? It doesn't matter if men want sex with thousands of beautiful and willing young women when in real life they end up having sex with maybe three women in their youth, and start a family with one. It also doesn't matter if a woman says she wants to marry a doctor from a wealthy family or stay single and date several different men until she's thirty-five. In terms of the evolution of a species, the only thing that counts is who actually has sex, conceives, and parents. In our human cultures, this translates into who marries and who has children.

SIMILARITY ATTRACTS

When married couples in this country are evaluated for their physical and emotional characteristics, an amusing result appears. People tend to mate with people who are of the same height, same weight, have similar personality traits, have about the same intelligence, and even the same nose width and earlobe length.[80] This commonalty starts at the level of dating, and it's from this pool of dates that we choose our marriage partners.[81] Even judged by outsiders looking at photographs of married couples, people marry someone of just about the same level of attractiveness.[82] And when asked to rate themselves, and then comment on who their best choice for a long-term partner might be, there's a surprising correlation between self-perception and what's important in a prospective mate. People basically mate to their own level, and would like a mate of similar intelligence, attractiveness, and emotional health.[83]

Everyone also tends to marry people that live in close proximity and share the same ethnic background, age, education, and religion.[84] And although men say they want to marry younger women, and women say they want older men, the actual age difference between men and women is only about three years or less, not particularly meaningful.[85] Also, over 90 percent of marriages in this country are made by people with the same racial background, although this must be changing, given the demographics of America.[86] Religion also has significant impact; over 60 percent marry others of the same religion. The issue of class is a mixed bag when it comes to marriage. Most people (58 percent) believe they married someone of the same social and economic class, and 22 percent think they moved up, while 20 percent say they married down. The same is true of education; most people marry those with nearly the same education, although a minority choose someone they perceive as smarter or dumber. These numbers translate across cultures as well. There are certain societies in which women are encouraged, or forced, to marry up, but most marriages occur between the same classes, religions, and ethnic groups.[87] And so, most often, like marries like.

And it goes even deeper. A group of geneticists collected blood from one thousand couples. They analyzed this blood by looking at ten blood proteins, often called blood types, such as ABO. This analysis is interesting because it tested for similarities or differences in blood types on real genes on six different chromosomes. They discovered that couples who have children are more genetically similar than any random pair of adults. These couples shared about 50 percent of their blood type markers in common, while random couples share 43 percent.[88] This doesn't mean that people are attracted to each other because they have B or O type blood. It probably means that, once again, we mate with those closest around us in our social circle who tend to be of our same race, class, and pool of genes.

It makes evolutionary sense to bump into, marry, and make children with someone like yourself. It means there are fewer areas of conflict, and if marriage and childrearing are a long-term venture, why not opt for as few conflicts as possible?[89] It's not someone's nose shape, pretty face, or tax record that makes marital navigation possible, it's sharing the same values and background. Everybody starts out by looking at the superficial, the face, the body, the clothes, but these are only clues to the person underneath.[90] What really counts are the things couples share in common. This isn't to say that marriages between opposites can't work; clearly they can and do. But most of us take the easiest route and marry someone who looks, acts, and comes from the same background as we do.

Marriage is a human universal. Divorce is also a human universal; people marry, divorce, and marry again in every culture.[91] People tend to form monogamous pair-bonds, although these bonds are not always exclusive, nor do they always last forever. Sex is a major feature of marriage, although sex with people other than spouses is also a human universal. We are attracted to some people and not others, we fall in love, or in lust, we make decisions about having or refusing to have sex. And those decisions are pushed by our deep biological urge to pass on genes as much as they are constrained by what society tells us we can and cannot do. We make a deal with ourselves and usually commit to one partner for a while. And we often bend to the rules our particular society sets out. And those rules, as we will see in the next chapter, have dire consequences for those who don't quite fit in.

CHAPTER SIX

The Natural History
of Homosexuality

"I would like to ask each and every straight person this question—when did you first know you were heterosexual?" The young man sitting across from me is Dimitri Moshoyannis, a student activist in the Gay and Lesbian Coalition at the university where I teach. He and fellow student Rebecca Rugg have volunteered to talk with me about the recent spate of research coming out in the news that points to a biological origin for homosexuality. Dimitri's question catches me by surprise—I am used to asking the questions, in a sense putting other people on the spot; for a moment I experience what it must feel like to be the person who's obliged to respond, to have the very core of my sexuality questioned by a stranger. How rude, I think to myself; this person I barely know dares to ask such an intimate question about my private life. In other words, his question makes just the point it's supposed to. It makes me uncomfortable and I'm offended. For a brief moment I also have some insight into what it must be like to be Dimitri or Rebecca and have my sexuality the topic of conversation.

This is admittedly an odd social time for human sexuality. We've had a sort of sexual revolution in this country a few decades ago and a bit of a backlash with the specter of AIDS shadowing casual sex. Homosexuality in particular is now the political football of the nineties. President Clinton began his tenure in office with a promise to end the ban on homosexuality in the military, only to waffle under the pressure of those who feared the consequences of allowing recruits and career military personnel to acknowledge their sexuality. In the fall of 1992, anti-gay legislation appeared on ballots in Oregon (defeated) and Colorado (passed). And everyone seems to have something to say about sexual orientation. Is it a personal choice? A lifestyle? Or a matter of inborn inheritance?

Science has played a major role in thrusting this issue into the public eye with several recent, and highly publicized, studies that suggest homosexuality has a biological basis. This kind of evidence is comforting to some because it implies that whom we love is not a matter of choice, but something hardwired from birth. Theoretically, people can't be punished or discriminated against for what is encoded on their genes. Others are leery of this research because it might further discrimination by opening the door to genetic testing or fetal testing, or might even place homosexuality into the category of a biological anomaly, marking this behavior as a pathology. And some might suggest that where there's a biological cause, there must also be a "cure." And so the drive toward figuring out if homosexuality is biologically based or not is a double-edged sword. On the one hand, it would be important to understand why a certain percentage of our species is more eroticized by their own sex than by the opposite sex. Understanding the biology of homosexuality would certainly help us also understand the biology, genetics, and development of heterosexuality. At the same time, any discoveries about homosexuality have political and social ramifications because our Western culture suffers from a long history of discrimination against and hate for homosexuals. Our historical attitude toward homosexuality echoes, I think, our troubled relationship with sexuality in general. In a sense, this recent focus on homosexuality in science and politics is self-defining; surely one reason we spend all this time and money searching for the biological roots of homosexuality is because we are so concerned, and uncomfortable, with our own sexual selves, no matter the direction.

What Is Homosexuality?

A man in prison for ten years masturbates a fellow inmate on a regular basis; once out of prison, he returns to his wife and has sex only with her. A thirteen-year-old girl kisses her best girlfriend and finds that she likes it. As part of a traditional ritual, an Anande boy allows an older male to have anal intercourse with him. A woman, married for fifteen years and the mother of two children, discovers she loves another woman the same way she loves her husband. Who exactly is homosexual here?

Humans have a driving need to categorize themselves and their behavior. We also like the world to be divided into nice, neat packages, because then we can affix labels, labels that define without question. But human sexuality, more than just about any human behavior, defies the strict rules of categories and labels. Homosexuality, for example, is a slippery term that wears many faces in any culture. And yet, Western society demands an explanation. Who is a homosexual? And more important to those concerned, *why* is someone a homosexual?

THE LABEL

Although some people in all cultures have, throughout human history, engaged in what we now call homosexual acts, the word "homosexual" is a relatively recent moniker.[1] Other terms, such as sexual inversion, uranism, sodomy, and pederasty, have been used, and most of them have referred to male sexuality involving other males. It wasn't until the early 1900s that the term "homosexual" was an accepted term for men and women who had sex with their own kind. Today, in the midst of what some might consider a homosexual revolution, the term "gay" is acceptable as a referent for male homosexuals and "lesbian" for women homosexuals. But it's still unclear, even with these labels, exactly what the labels mean.

Part of the problem is that as a society, we are confused about the word "sex." First, each of us has a sex, or biological status. Second, people are defined culturally by another kind of sex, their cultural role. And third, there's sex the act, which might have nothing to do with the other two kinds of sex. The biological sex of a person is usually defined by the products of reproduction; women as a sex make eggs and men as a sex make sperm. Some individuals have abnormal complements of chromosomes or enzyme deficiencies and make neither eggs nor sperm, and in a sense, are an ambivalent sex.[2] The cultural role of sex is probably the kind of sex we most often deal with day to day. It involves the culturally constructed, and sanctioned, outward appearance of one sex or another.[3] Today, we often call this "gender" rather than sex to underscore that this aspect of "sex" is cultural, variable, and thus highly flexible rather than set by biology. As psychologist Sandra Lipsitz Bem points out, this layer of sex in our culture and most other cultures usually establishes a polarization into male and female defined by dress, social roles, and acceptable personality styles. For example, in Western society, men most often wear pants while women wear dresses; men perform manual labor such as construction work while women care for children or do desk work; men handle the barbecue and women run the dishwasher. These gender designations are reasonably flexible, and they change with time or revolution, but they are constricting just the same. Most often, if someone doesn't quite fit within the confines of the societal bounds of one gender or another, they're considered unnatural, immoral, or abnormal biologically or psychologically.[4] Set within the biological designation of sex and its genderized cultural accessories is the act of sex, which isn't really just an act at all. Sex the behavior, as I've made clear in this book, is a matter of biology, culture, and psychology. It's a process that involves thought, desire, and body movement over a span of time. There are all sorts of ways to engage in sexual acts, but they usually involve genitalia and orgasm. There are, of course myriad ways to achieve the end of sexual release.

And so our confusion about who exactly is a homosexual stems from which kind of sex we are talking about. Is a man who likes staying home with his kids a homosexual? Is the woman who wears pants a homosexual? Is the man who wears a dress a homosexual? No. Homosexuality is most often, in our culture, defined by a sexual act that occurs with someone of the same sex. In that case, is the man who had anal intercourse once in his teens a homosexual? Is the woman who lives with another woman but every once in a while fantasizes about men during sex a homosexual? This basis for definition is just as confusing as one relying on gender. The label, in the end, has to be one of self-definition. And that definition, some feel, is a whispering within that each one of us listens to alone, a definition that society cannot alter. John Money, who has spent his life studying the endocrinological basis of sexuality, believes we are defined sexually by whom we fall in love with. According to this definition, people who regularly fall in love with someone of the same sex are homosexuals, no matter their products of reproduction, their gender, or if they ever actually act on their desires.

HOW MANY PEOPLE ARE HOMOSEXUALS?

Most often, a figure of 10 percent of the total population for male homosexuality and half that number for female homosexuality has been quoted across cultures.[5] This number isn't based on empirical data or extensive interviews; it's just an estimate. Recently, the figure of 10 percent has come under fire. Surveys in the United States, Canada, Britain, and France report a range from 1 percent to 7 percent of the total population for those who consider themselves exclusively homosexual.[6] These and other studies have reported that although many people may have had one or a few homosexual experiences in their lifetime, only a small percentage of the population consider themselves exclusively homosexual.[7] Although there are few statistics available from other countries (see below), the figures that are available tend to be even lower. However, these numbers are systematically unreliable because of the methodology used in gathering the data.[8] All of these studies have been conducted, and are publicized, because of the political ramifications of this particular numbers game. If homosexuality is "common," homosexuals have a strong case for discrimination against a large portion of the population and more clout in fighting for social justice. If, on the other hand, the numbers are low, they might not have enough power in numbers to constitute a strong minority voice. And so the numbers count.

The most extensive study, and the one quoted most often, is Kinsey and his colleagues' work from the 1940s in which they surveyed 5,300 men about their sexuality.[9] Faced with the same definition problems that I outlined above, these researchers decided to stick with an orgasmic definition of homosexual acts. For the purpose of their research, they defined any orgasm—self-masturbation, mutual masturbation, oral or anal sex—that occurred in the presence of another man as a homosexual act. They determined, through personal interviews and questionnaires, that 6.3 percent of all orgasms by their study population had occurred in the company of other men. In addition, they found that 37 percent of their sample had had at least one orgasm with another male in his lifetime. Most of these interactions had occurred during adolescence; 60 percent of those men admitting to a homosexual act said it happened during their teen years. These data, as Kinsey pointed out, were not so much a measure of how many homosexuals there were in America, but how many men had had a homosexual experience.

Kinsey was especially offended by any pigeonholing of male or female sexuality into polarized categories of homosexual or heterosexual. Instead, he devised a scale of sexuality from 0 to 6. Under his scheme, those who only have thoughts, fantasy, and sex with the opposite sex rate a 0—forever and exclusively heterosexual. Those who say they are, and have always been, homosexual in thought and action rate a 6. In between are people who think about, and often have, sex with either sex. We label these people as bisexual. Kinsey felt that there were few people on either extreme of the scale; most heterosexual people, he felt, at least have had a single homosexual fantasy in their lifetime.[10] Rating his male subjects by this scale, 13 percent were moderately homosexual and only 4 percent were exclusively homosexual. Interestingly enough, this 4 percent figure is most often cited by more recent studies.[11]

The numbers for female homosexuals are even less clear. While most authors state that female homosexuality occurs at a much lower rate than male homosexuality, no one really knows what those numbers might be. There are various reasons why any numbers on female sexuality, and this includes lesbianism, are suspect. We lack adequate numbers on cross-cultural and Western female sexuality because, until recently, researchers haven't considered female sexuality much of an issue; the data aren't there because no one asked. And even when women are asked, they have been taught in most societies that sex is a taboo subject that shouldn't be discussed, especially with strangers. So even when asked, researchers might not have been told the truth. The numbers on lesbians in particular are probably inaccurate because in all societies, female sexuality is oppressed, and women with homosexual tendencies may be forced against their natural inclinations into the traditional roles of wife and mother. This might explain why lesbians tend to acknowledge their sexual orientation later than men, and often marry or have long-term heterosexual relationships before coming out of the closet. In addition, some lesbians claim that they have consciously chosen the lesbian life as a political statement against male oppression, which complicates a count of "natural" lesbians. And finally, lesbianism doesn't offend our Western culture's sense of male machismo. As a result, society and scientists probably ignore female homosexuality and don't think too hard, or care too much, about the numbers involved.

No one really knows how many people have experienced homosexual acts, how many people actually consider themselves homosexual or bisexual. And surely no one will ever know how many heterosexuals have wondered what it might be like to make love to someone of the same sex. We can only be sure that in every human group, some people, sometimes or all the time, have sex with the same sex, or have thought about it. That is, homosexuality is a human universal.[12]

HOMOSEXUALITY ACROSS CULTURES

There's never been a systematic study comparing homosexuality across cultures. Instead, anthropologists have gathered information in a few societies in which homosexuality is an obvious part of the community. As a result, we really don't know the percentage of humans across the globe who are homosexual. This information is important because it would support, or detract from, either the biological or psychological origin theories of homosexuality. If the same rate of homosexuality appears in all cultures, say around 4 percent for men and 2 percent for women (which is now used for Western cultures), this would confirm the argument that no matter the individual psychology, no matter the cultural pressures, some members of our tribe will always be homosexual. That is, homosexuality must be a biological feature of our species. If, on the other hand, the number of homosexuals fluctuates among cultures, or isn't found in some, then a psychological or cultural explanation seems more appropriate.

With the little cross-cultural data available, it appears that homosexuality, that is the sexual act among same-sexed partners, is common in many cultures. There are, however, a few groups who claim they have never heard of such a thing.[13] One anthropological database called the Human Relations Area File (HRAF), which catalogs accounts of societies written by anthropologists, includes information, that is ethnographic descriptions, on about two hundred societies around the world. In an analysis of cross-cultural homosexuality, researchers found that there was information on male-male sexuality in seventy-six societies.[14] Female homosexual behavior was reported in only seventeen of these same seventy-six societies.[15] This isn't to say that homosexuality doesn't occur in the other one hundred twenty-four, only that the subject just didn't come up during the ethnographer's stay. Of more interest is the attitude of those seventy-six societies to homosexuality. In 64 percent of these groups, homosexual behavior is either acceptable, tolerated, or considered normal. In other words, there is no label of "deviance" to the act. In some groups, homosexuality is tied up with various layers of the social fabric such as status and economics.[16]

There seems to be an overall association between how a society catalogs people by gender identity and their reaction to homosexuality. Disapproving societies tend to be highly gender-polarized, and they punish those who even dare to cross gender lines. For example, the Mexican mestizo culture values maleness—males have high status and power while women have low status and little power. Men who don't conform to an extreme machismo ideal are pushed in the opposite direction into the feminine role that has also been culturally constructed; they are assumed to be passive persons of low status and treated as such. Homosexuality in this culture is a term reserved for the "feminized" males who must always assume the female, and what this culture sees as passive, position of insertee during anal intercourse. Men who insert during what our culture would count as a homosexual interaction are not considered "homosexual" because they aren't acting the passive-female role during sex and therefore maintain the machismo dominance. The receivers are stigmatized and ridiculed while the inserter is ignored.[17] The distinction between who does what to whom during male-male sex and who gets stigmatized as homosexual is found in similar machismo-oriented cultures in Greece and Turkey.

Other cultures sometimes have a place for those who don't conform to particular gender roles. For example, several Native American cultures, before white contact, included men called berdaches.[18] In boyhood, these unusual males were recognized as different, and allowed, even encouraged, to take on the traditional female role in dress and occupation. The berdache was not considered deviant, but a functional member of the society. Often, the berdache was regarded as a man of power, a shaman, and his special role and sexuality were believed to be a result of a supernatural transformation. What makes the berdache different from a gay male today in these same societies is that the tradition of berdache, that is taking on a female role as well as being oriented to male-male sexuality, has died out, but homosexual orientation continues.[19] In northern India, a group of men, including eunuchs, transvestites, transsexuals, hermaphrodites, and homosexuals, are organized into a religious community and called hijras. They are viewed as neither male nor female, but as a sex in between, an alternate gender.[20] These men undergo castration, but this operation renders them otherworldly, they believe, because they are impotent as men, but unable to reproduce as women. Some hijras have relationships with non-hijras men which they consider marriage. Their role in the Indian social fabric is to bring blessings and perform dances at ceremonies and rites of passage. They do, however, sometimes encounter disapproval and ridicule. They are not isolated from society and have a certain amount of religious power, but nonetheless, they don't exactly fit in either. In any case, the hijra is an established, accepted gender role India. Men in acceptable female roles have been observed in such geographically diverse cultures as the Chukchee of Siberia and the Tanala of Madagascar.[21] Some have suggested that our Western view of gender, which is strict, makes us unable to accept the broad range of gender roles that appear in other cultures.[22]

Homosexuality is also tolerated in societies where males are separated from females and confined together. For the Ngonde of Africa, homosexual acts between boys before they marry are fine as long as it's mutually agreeable to both parties.[23] We can probably assume this kind of tolerance often happens in our own culture in prisons or boys' schools where opportunity and restriction make odd bedfellows.

In other cultures, anal intercourse or some form of male homoeroticism is considered necessary for manhood.[24] Among the Aranda of Australia, unmarried men will take on a young boy and live with him, and have sex with him, until the older man marries. For the Keraki of New Guinea, anal intercourse is part of the puberty rite of every boy. At first, a Keraki boy must receive the favors of other men, but after a year, he penetrates smaller boys on their way to manhood. The Sambia, another New Guinea group, practice obligatory fellatio for all males. Boys are raised by their mothers until they're seven years old. At that point, they join a cult of older boys and unmarried men and are instructed to fellate their elders and ingest semen every day. Once small boys reach puberty, smaller boys fellate them. All males engage in this exchange of semen until they marry and start to focus on sex with women; homosexual activity stops at the birth of the first child. This institutionalized passing of semen is essential among the Sambia because they believe semen isn't a natural product of male biology but something a boy must get from other men to grow strong and move away from femaleness, the more original state.[25] All this semen passing takes place in an atmosphere of tense and hostile male-female relationships; boys are taught to fear women and think of their bodies as taboo. Beyond the sexual pleasure that might occur among the men and boys there is clearly a male bonding effect of their homosexual interactions which has significant social consequences. Although some in Western cultures might find these practices disgusting, we have our own homoeroticism to own up to. Mutual masturbation among adolescent boys has been reported as a regular feature of Western culture, and yet it's uncommon in other cultures.[26]

Information on female homosexuality is more scarce. In one group, the Aranda of Australia again, women might rub the clitoris of another woman for pleasure. Chuckee women sometimes employed an artificial penis made of the large calf muscle of a reindeer and used it on other women. Use of objects to rub each other is noted in other groups as well. One recent study of homosexuality in Mombasa, a town on the Kenya coast, describes a number of lesbian relationships that form an acceptable subculture, even in this primarily Muslim community.[27]

Although these descriptions sound, to our conservative Western ears, like anecdotes out of *Ripley's Believe It or Not,* I use them for a reason. My point is that the industrialized West shares an attitude toward homosexuality, a negative and disapproving attitude, that is molded by tradition, politics, and religion. Specific to our Judeo-Christian ethic is a mandate that declares any sexual interaction that cannot possibly result in conception an act against Nature. Thus homosexuality, in any form, is unnatural, and by implication, perverse.[28] We also live in a world of very polarized gender identification. Men must look and act a certain way and women should know their place too. And so we have trouble when gender and sexuality don't conform to the notions we've learned in our culture from childhood. And this history makes as much sense in our world as a more permissive and encouraging stance toward homosexual acts does in other cultures. The most important point to learn from this catalog of human behavior is that sexuality, and this includes homosexuality, is a highly flexible pattern of behavior in our species. Any person *can* engage in what we have decided to call homosexual acts. The question is why some people do, while others do not. And more important—

why are some people compelled to mate with the same sex because this is the only body type that fills them with desire?

What Makes a Homosexual?

As noted above, the term "homosexual" is best defined by the acts that are performed. Some people seem confused, and often disgusted, by what homosexuals do. Well, homosexuals have sex just like heterosexuals. The only difference is the anatomy of the partner at hand. Homosexual sex involves kissing, touching, genital manipulation, oral sex, vaginal sex, and anal sex.[29] Choose a type of partner and then see how it works out. For those who have experienced a homosexual interaction, or are primarily homosexual in their love lives, this way of sex is perfectly natural. For those who have never experienced sex with a same-sex partner, the pattern seems alien. I recently came up with a personal analogy for my thoughts on homosexuality. I am heterosexual and I have no children. The thought of being a lesbian is as alien to me as the thought of being a mother. In both cases, I see people all around me in those roles, as lesbians and as mothers, and I see advantages and costs to both. But most important, I am not quite sure how it feels to be a lesbian or a mother—I understand each role, and I certainly feel no disapproval, but the experience of each is completely alien to me, and yet we are all women. These are just two levels of femaleness that are not open to me, and so I will never fully know what it is like to be either a lesbian or a mother—it's just not in the cards. There are other roles, feelings, and ways of life that are also alien to me. I can't imagine being a man, or being rich, or being homeless—all common human experiences, but ones I will probably never have. And so when someone says to me, "I just don't get it. How can a man love a man, or kiss a man?" my response is, "I also don't 'get' homosexuality. I don't 'get' it in the same way I don't 'get' what it's like to be a man, rich, homeless, lesbian, or a mother." Oddly enough, although there are so many human experiences that are open only to some and not others, our culture has spent many years, millions of dollars, and much thought trying to figure out exactly why someone is a homosexual.

THE PSYCHOLOGY OF HOMOSEXUALITY

Before the sexual revolution and gay liberation, most people assumed that homosexuality was a product of a twisted childhood. This approach was fostered by the field of psychiatry which had its roots in Sigmund Freud. But Freud alone can't be blamed for this. Homosexuality, Freud believed, is not a pathology or an illness.[30] Freud did believe, however, that all people are born bisexual and with maturity, each one of us moves from these childish ways and becomes more interested in the opposite sex. And because he saw sexual development as a process, it followed that a person could also become arrested at the homosexual stage and never mature into "normal" heterosexuality. Psychoanalysts after Freud soon wove a scenario for what they saw as a "typical" male homosexual background based on Freud's concept of the Oedipal conflict. The scenario includes a clinging, hostile mother who restricts the behavior and emotional health of her male child, and a father who is aloof and uncaring. The mother is at fault because she doesn't allow the boy to mature and disconnect from her emotionally. The father is at fault because he takes no steps to help this boy detach from his overbearing mother. As a result, the boy feels rage, and repulsion, for all women, and is therefore more drawn to men.[31] This theory was presumably substantiated in the early 1960s by an extensive study of 106 homosexual men.[32] These men, it seems, did have unhappy family lives and were undergoing treatment with 77 different analysts. According to a battery of psychological tests, the subjects appeared to have a hidden fear of women. Homosexuality, in this view, is a result of turning away from one sex in horror, rather than being positively oriented toward the same sex.

And so male homosexuality, according to this view, is a result of a pathologically abnormal triangle of an overbearing mother, an aloof father, and a child who reacts to their strangeness. By implication, it follows that homosexuality is psychologically driven, not a product of anything biological, and thus open to "cure" when the underlying neurosis is addressed. In all these discussions, the psychological cause of female homosexuality seemed to be of little interest. In 1952, the American Psychiatric Association was so convinced of this scenario that in the first edition of their manual used to standardize mental conditions, *The Diagnostic and Statistical Manual* (DSM), homosexuality was listed as a sociopathic personality disturbance, a mental illness.[33] Over the next twenty years, a number of psychiatrists lobbied to remove homosexuality from the DSM. They knew that the psychoanalytic view of homosexuals came from sessions with men who were in therapy because they were unhappy with their lives, or uncomfortable with their sexuality. A better model, suggested by analysts such as Judd Marmor and Richard Isay, would be modeled after emotionally healthy homosexuals and the various stories of their upbringings, rather than unhealthy men who happen to be homosexual.[34] Instrumental in this new approach was work by psychiatrist Evelyn Hooker. In the 1950s Hooker began studying sixty men and their psychological histories. Thirty of these men were professed homosexuals, and the other thirty were heterosexual. None of the men had sought psychiatric care ever in their lives. Using a battery of standard psychological tests, Hooker found that there was no difference in the mental health of these men; the homosexuals were no more well or badly adjusted than the heterosexuals.[35] This work also showed that male homosexuals who were comfortable with themselves, that is men not seeking therapy because of their homosexuality or because they had a generally unhappy life, have the same variety of family experiences as heterosexual people. In other words, oppressive mothers and aloof fathers are no more common among homosexual men than among heterosexuals. This view states that upbringing and society can certainly influence the expression of anyone's sexuality, but in most cases the natural love object is a given.[36] By 1968, these therapists were successful in moving homosexuality from the classification of sociopathic disorder to a new category, sexual deviance. It wasn't until 1973, and under great protest, that the APA removed homosexuality altogether from the DSM.

Today, although most therapists accept that homosexuality is usually not a psychological response to bad upbringing, there's still a certain contingent of psychoanalysts that adhere to the overbearing mother–aloof father model, and some of these analysts report success in "normalizing" their patients.[37] This normalization allows homosexual men to lead heterosexual lives, but it doesn't seem to wipe out homosexual desires completely, as their patients admit. And so psychoanalysts and psychologists are often left searching for some other explanations for homosexuality beyond the psychological.

CHOICE OR FATE?

As in the membership of the APA, there are still many people in Western culture who feel strongly that homosexuality is a pathology, mental illness, or a sin against Nature. At issue is the root of this aberrant behavior —is it fate, or a chosen path? If psychology is not pushing a person toward homosexuality, some think, they must be making a conscious choice, a "lifestyle choice" to be homosexual. Their view is based on the assumption that a person can physically and psychologically choose what turns them on and whom they fall in love with. And in a sense this is correct. We do have some say in whom we mate with and marry. But it's highly unlikely that we can consciously control our sexual impulses toward one sex or the other; falling in love is exactly that, a fall, not a conscious step. This more fatalistic viewpoint has gained support not only because it seems to make common sense, but also because of a flock of studies published in the past two years in the scientific and popular press that point to a more biological origin for human sexuality.

The Biology of Homosexuality

There are several ways a trait or pattern of behavior can be labeled "biological" in origin. When I ask the question Is there a biological basis to homosexuality? I am actually asking several questions. The biology of an organism includes a discussion of not only genes, but also hormones, cell growth, physiology, and any organic changes that might occur. It also includes enzymes, cell division, aging, growth, and cell mutation. In other words, biology is not just genes. More important, none of these "biological" factors stand alone, not even genes.

For example, each person has genes from his or her mother and father that combine and produce a certain amount of melanin in the iris of the eye. That set program of melanin, or lack of it, makes the iris appear blue, brown, or something else. The color is permanent. I assume no one would debate the statement that eye color is a genetically determined trait. Height is also a biologically determined trait, but not in the same way. Combined genes from both parents determine if a person will be short or tall, but how much a person eats and what kind of life they lead also influence their ultimate height. For example, two short people can give birth to a short child who turns into a basketball star because he avoided childhood diseases and ate well. This trait is genetically determined but also greatly influenced by the environment. An organism is born, grows, and functions in an environment. That environment for humans includes the air we breathe, the food we drink, the family we grow up in, and the culture we belong to. Although some would like to separate biology from environment, we can't. And so the first point to remember about the biology of any behavioral pattern in humans, and other animals for that matter, is that the label "biological" origin doesn't necessarily mean that the environment has no role to play in the ultimate expression of any kind of behavior. This is especially true for sexuality. Patterns of behavior, especially all the thousands of reactions, stipulations, actions, and feelings involved in sexuality, are not single traits such as eye color and blood type. This should be an obvious point—I have described sexuality as a highly variable continuum, and by definition, no single gene or any set of genes could account for the prism of a sexual continuum.

Our biology and our environment are too entwined to pull apart. Instead, the exercise is one of figuring out the biological mechanism that is involved in the expression of any particular behavior or trait. We waste our time looking for a strict biological cause for anything as complex, as life-enduring, and as flexible as sexuality. But we don't waste our time looking for the biological elements to our sexual lives.

The search for a biological basis for homosexuality in particular has taken many paths, including research on genetics, neuroanatomy, hormones, and twin studies. Scientists assume first and foremost that some kind of genetic influence is at work. These special genes might direct the development of areas in the brain, or produce hormones of sexuality that could eventually orient someone to be attracted to a same-sex partner. On the surface, this kind of physiological reasoning makes sense. But a closer look at all the research to date on the biology of human homosexuality falls short of telling us why our brothers, sisters, children, and parents turn out straight or gay.

THE GENETICS OF HOMOSEXUALITY

So far, there's no such thing as a gene for homosexuality. This also means a gene for heterosexuality hasn't been identified either. But there is provocative news about an area of the X chromosome that might lead to the discovery of a set of genes that somehow predisposes the direction of love.

In the summer of 1993, Dean Hamer and his research team at the National Institutes of Health published a study that looked at the inheritance pattern and chromosomal topography of gay men.[38] The team traced the family histories of seventy-six self-defined male homosexuals. These gay men had a significantly higher number of homosexual brothers (13.5 percent) than heterosexual male controls (2 percent). This familial similarity between homosexual male siblings, or concordance as behavioral genetics call it when two individuals are the same, had been found in other studies as well.[39] More significant, the researchers mapped out the extended families of these men as far as they could. Surprisingly, the distribution of male homosexuality was not scattered randomly across family trees. When the families of the homosexual subjects or "probands" were traced over a number of generations and out to cousins, a significant number of maternal relatives, that is the mother's brothers and the sons of the mother's sisters, were identified as gay. The consistent pattern in these pedigrees suggested not only that homosexuality has a genetic and inherited component, but that it also follows the maternal line. There are, of course, problems with this kind of pedigree analysis. Gathering information on generations of families is hit or miss at best. And trying to identify a great-granddad's sexuality is next to impossible. But the fact remains that when the researchers were able to tag gay men, they most often were related on their mother's side, and paternal uncles and cousins came up heterosexual.

Hamer's groups didn't stop there. Since men inherit their only X chromosome from their mother, the researchers decided to focus specifically on the X chromosome. Blood samples from a subset of forty pairs of homosexual brothers were collected so that the researchers could genetically compare the X chromosome of one homosexual brother to another. They used chromosomal markers, repeated sequences of DNA that can be easily isolated, tagged, and identified. Markers are not real genes, but a way to catalog strips of gene sequences. Think of it this way: On the X chromosome is a box full of various strings of Christmas lights; some have only red bulbs, some have only white bulbs, and they are all tangled together so that you can't pick out any particular kind. We want to find only the red strands, and so we locate plugs that indicate red strands and plug them into a socket, and *voilà*, the red strands are marked for easy untangling. Hamer and his colleagues used twenty-two known markers running the length of the X chromosome (think of twenty-two different possibilities of colored strands of light) and compared the type of markers across the pairs. Just by chance, because these men were brothers, they should share 50 percent of their markers in common. Taking this into account, the brothers shared or didn't share about the right number of markers at most locations on the X chromosome. But the lower tip of the X chromosome, an area known as Xq28 in genome-speak, was different. Thirty-three of the forty pairs of gay brothers under study shared an identical pattern of five DNA markers in this area alone. In other words, the markers in this one section of DNA were shared by just about all of them. The key to the genetics of homosexuality might, therefore, lie on the tip of the X chromosome, and be passed from mothers to sons.

For those who believe in a genetic foundation for homosexuality, this might seem like the answer. But it's not. Even Hamer admits they've only identified a region of the X chromosome that seems to most often look the same in brothers who are gay; this isn't the same as finding the genes for homosexuality. This region contains about 4 million base pairs and represents less than 0.2 percent of the human genome.[40] Finding a particular gene, or a set of genes, that might direct sexuality in this area is the proverbial needle in the haystack. And even if they do find particular segments of DNA that account for sexual orientation, that isn't the final answer either. Several families showed a paternal (through the father) rather than maternal pattern of homosexuality, so that the assumption of gay genes on the X chromosome wouldn't work for them. Also, seven of the brothers in this study didn't even share the same markers. If these are the supposed genes or markers for homosexuality, why are they absent in some? And finally no one, not even Hamer, has any idea exactly what these markers code for. It's possible they code for a certain brain chemistry or physiology that eventually translates into a type of sexual orientation. But for all we know, the approximately one hundred genes in this area code for something other than anything to do with sexuality. And so the supposed genes for homosexuality, or heterosexuality for that matter, remain to be discovered.

A TALE OF TWINS

Before the days of sophisticated genetic analysis, biologists relied on studies of twins to figure out how much of human behavior might be genetic. Identical twins, called monozygotic twins because they're conceived by one egg and one sperm that divides into two individuals, have the same genes. But from the day of birth, each twin reacts to the world a bit differently. Scientists speculate that any traits or behaviors that these twins share in common are highly genetic, while those in which they differ must be due to environment. Fraternal twins, called dizygotic twins because they are conceived with two eggs and two different sperm, are no more genetically alike than normal siblings, but they did share the same fetal environment for nine months. These various twin sets represent a natural experiment for scientists interested in determining the genetic basis of behaviors.

Several studies conducted over the past forty years have shown that if one twin, either monozygotic or dizygotic, is homosexual, the probability of the other twin being homosexual is higher than if they were regular siblings.[41] Even when twins are reared apart, there seems to be some pure genetic connection to their sexual orientation.[42] These studies are flawed, however, because they relied on populations of mentally ill patients, or the number of subjects involved were extremely small. Recently, Michael Bailey and Richard Pillard used a relatively large sample of men and women twin sets, people recruited from the country at large, to hunt for a genetic component for homosexuality.[43] Their first study of homosexual men included one hundred and fifteen twin sets. Some were classified as monozygotic twins, that is having identical genes, while others were classified as dizygotic twins because they looked different enough to suggest that they had had different genes. The sample also included forty-six gay men with adoptive brothers. In other words, there were gay men who shared both their exact genes and fetal environment with their twin brothers, gay men who had the same fetal environment as their brothers but shared only half their genes, and gay men who shared the same family environment but had absolutely no genes in common with their adoptive brothers. Of the fifty-six sets of identical twins, one of which was homosexual, 52 percent had a twin brother who was also homosexual. When the twins were not identical, only 22 percent said their brother was also homosexual. Both rates are far above the national average of about 4 to 10 percent for homosexuality, suggesting that fetal environment as well as genes play an important role in guiding sexual development. Oddly enough, adoptive brothers were also gay in 11 percent of the cases, while nontwin siblings were gay in only 9.2 percent of the cases. In addition, Bailey and Pillard have recently found similar results for female homosexual twins, with monozygotic twins sharing the highest concordance for lesbianism, nontwin sisters the least, and dizygotic twins somewhere in the middle.[44] The finding was recently verified with sixty-one pairs of male homosexual twins in another study with the same methodology. More interesting, this other study included three sets of triplets in which at least one triplet was homosexual.[45] One set included a pair of monozygotic brothers who were both homosexual and their heterosexual, dizygotic sister. The second set consisted of two lesbian monozygotic sisters and one dizygotic heterosexual sister. And the third set were monozygotic homosexual brothers. If anyone doubts the power of genes in directing homosexuality, these triplet sets ought make them pause and think.

What's important in all of these studies, according to Bailey and Pillard, is the marked difference between identical and nonidentical twins. When twins share the same genes, and one is homosexual, the co-twin is twice as likely to become homosexual than if he had merely shared the same fetal environment. Genes, they conclude, play a major role in determining sexual orientation, and thus homosexuality is moderately heritable. This means there might be a genetic component, maybe even a predisposition, to homosexuality, but environment also plays a large role in determining if the homosexual tendency is eventually expressed. Homosexuality, in their view, is not genetically determined, but certainly genetically influenced.

Bailey and Pillard's work, and other twin studies, have come under fire. Since the twins in this work were raised in the same household, the study can't reasonably factor out genes from environment. These studies also rely on questionnaires or the opinion of family members to label a person's sexuality, and it's hard to think this is reliable data of someone else's sexual experience or desires. More basic to the science of these twin studies is what little we know about the genetics of twinning. Monozygotic twins may, in fact, not share every strand of DNA in common.[46] The variability of what genes twins carry might account for the fact that the rates of concordance, that is both twins being homosexual, vary across studies from a few percent to 100 percent. More striking, points out a critic of this work, William Byne, is the high number of identical twins that are *not* concordant for homosexuality—almost half— suggesting that many other factors besides genes guide the sexual orientation of an adult.[47] Like all other human behaviors that geneticists have looked at, say schizophrenia or alcoholism, genes can only explain so much. And so the twin studies suggest, but certainly don't conclusively prove, a genetic component to sexual orientation, be it gay or straight.

IT MUST BE HORMONAL

If we accept for a moment that genes are somehow implicated in sexual orientation, how might this work? The most obvious place to search for a biochemical mechanism for homosexuality is in hormonal action, which is regulated by the brain and dictated by genes. Hormonal profiles, then, might demonstrate differences in sexual orientation as well as differences between men and women. Hormones are powerful forces in our behavioral and physical development, and they are directly related to sex and reproduction. The basis of this research asserts that during fetal development, and for a few months after birth, the brain is the traffic director of character development. Apparently, there are critical periods when a genetically male fetus becomes awash in hormones that independently masculinize and defeminize the brain—these hormones "organize" the brain for later life as "male" or "female."[48] It follows, therefore, that if something goes wrong during these critical periods, times not yet clearly defined, the brain might end up incomplete in one direction or the other. For example, an incompletely bathed male brain might develop into an adult sexually excited by other males because the prenatal brain was not fully "masculinized."[49]

Much of the work on hormones and their role in female or male sexuality is based on brain studies of rats.[50] When male rats are castrated soon after birth, eliminating the male hormone testosterone at a critical period when the brain is becoming masculinized, they regularly display the female mating position, called lordosis. In lordosis, an animal crouches low, stays still, and rears its hind end ready for copulation. Female rats exposed to androgens early in fetal life, which should have a masculinizing effect on the brain, show the more male-typical mounting behavior as adults.[51] They jump on the backs of other females and hump them as a copulating male would. The only problem is that the changes in sexual behavior of these neurally and hormonally mixed-up animals are neither consistent nor absolute. The lordosis-oriented males, for example, continue to mount other males, suggesting they still retain the more "male" pattern of sexuality along with the "female" acquired pattern.[52] The appearance of lordosis in males is a shaky indicator of changed sexuality anyway. Lordosis is a behavior exhibited during both sexuality with other rodents and handling by human caretakers; it's a response to a physical stimulus, not just a feeling of sexual excitement. Also, hormonally normal males mount males in fake lordosis, raising the question Who exactly is acting in a homosexual manner here? More important is the relevance of these rodent studies to human behavior. Rodents are stimulus-response animals and their mating behavior is under strict hormonal control. Unlike rats, humans don't respond uniformly to every sexual stimulus; our sexuality can't be manipulated quite so easily as the sexual response of our rodent friends.[53] But the rodent work does propose a developmental process that involves two independent steps of masculinization and defeminization for male fetuses, while female fetuses follow a default hormonal pathway to a female brain. For humans, this might mean that high levels of male hormones early in fetal life could result in hormonally heterosexual males and homosexual females—they will both have brains sexually organized in a male direction, and love women. Low levels of male hormones in fetal life will presumably result in male homosexuals and female heterosexuals who have the sexual brains of females and are attracted to men.[54]

The vagaries of human physiology have presented us with a ready-made human laboratory for analyzing the role of prenatal hormones in sexual orientation. There are children born with male genes, but lacking the receptors for male androgens. There are also genetic females who have, for various reasons, been exposed to large amounts of male hormones during their fetal development. These people are living examples of the effects of hormone alterations on human sexuality. Clearly, hormone levels early in fetal life that are incompatible with the sexual genetics of a fetus result in dramatic alterations in internal reproductive organs and external genitalia. For example, a female fetus, that is a fetus with an XX chromosomal complement, who is exposed to too much androgen will have masculinized genitalia and be what physiologists call a pseudohermaphrodite.[55] The genitals of these girls are often surgically altered to make them completely "female." Psychological profiles show that these girls often express tomboyish behavior during childhood and some become homosexual. But no one knows if their prenatal hormones, their medical experience, their upbringing, or their odd place in our culture is responsible for their sometimes "masculine" behavior.

Another group, genetically XY males who suffer from testicular feminization, have testes but do not respond to testosterone. These boys develop female-like breasts and their testes don't descend into scrotal sacs or pseudo-labia. They are usually raised as girls and learn a female sexual role, regardless of their chromosomal, anatomical, or hormonal designation.[56] A more curious case concerns a disorder called 5-alpha reductase deficiency. Males born with this syndrome have remnants of both male and female genitals and are most often called, and raised as, females. At puberty, these "girls" suddenly masculinize in that their voices drop, their testes descend, and their tiny penises enlarge. A study of this disorder in one family in the Dominican Republic tracked a corresponding gender role transformation by some of these pseudohermaphrodites.[57] Some individuals in the study stayed "female" at puberty while others became "male." If hormones were driving the sexual motivation of these individuals, it seems they should all transform into males or remain females. Instead, there is no typical pattern. In other words, even when Nature missteps and produces people with combinations of hormones and sexual anatomy, there's no straight path to maleness or femaleness, and sexual orientation in these cases is certainly not driven by hormones alone.[58]

A lack of a hormonal basis for sexual orientation also appears in work on the endocrinology of homosexuals.[59] There is no difference in the levels of testosterone between male heterosexuals and male homosexuals, or in the levels of estrogen in female homosexuals and female heterosexuals, and both sexes have the correct amounts of gonadotropins regardless of their sexual orientation.[60] And injections of testosterone have no effect on male homosexuals other than to increase their sex drive—for other men.[61] Women treated with powerful hormone therapy during pregnancy produce no more homosexual children than other women. And the possibility that stress might alter a woman's body chemistry and push her growing fetus toward gayness has no support whatever. Mothers of homosexuals don't remember their pregnancies as more stressful than mothers of heterosexuals.[62] In fact, there's no information on women under clear stress, say during wars or famine, and the eventual sexual orientation of their children.[63] In addition, women experience pregnancies under a vast array of conditions—different amounts of nutritional, emotional, and financial stress—but the rate of homosexuality remains the same.

Attempts to manipulate hormonal pathways and prove that male homosexuals at least respond to hormones differently, and are therefore endocrinologically female, have also been inconclusive. In one study, twenty-one homosexual men were injected with 20 milligrams of Premarin, a synthetic female estrogen. The researchers expected these men to react hormonally the way a woman would—a decrease in levels of leutinizing hormone followed by a sharp rise, a pattern which typically occurs at ovulation. The homosexual men did react this way, but at a much slower pace, while the heterosexual and bisexual male controls showed a decrease in LH level, but little rise in LH.[64] In another study, conducted ten years later, homosexual men responded to jolts of estrogen in a manner intermediate to women and heterosexual men.[65] And although all the male subjects had lower levels of testosterone after injections of estrogen, the heterosexual males quickly regained normal levels of circulating testosterone while homosexuals took much longer. One researcher believes this "homosexual" response to estrogen is caused by insufficient amounts of prenatal testosterone, which in turn results in an incompletely masculinized hypothalamus and a more feminized pituitary.

The basis for this hypothesis, and the reason for these studies in the first place, is inappropriate. Recent evidence shows that the LH-estrogen feedback mechanism is not really sexually dimorphic in the first place in any primate, including humans.[66] Therefore, it's silly to look for a similarity between homosexual men and heterosexual men when we don't even know if there's a real difference between men and women. More at issue is what this fabricated hormonal response is supposed to mean in the real world. Injections of Premarin and the responses they invoke have nothing to do with the prenatal brain that is supposedly forming a particular sexual direction.[67] Most important is the criticism that LH response in men is really regulated by testicular function, and testicular function is easily affected by alcohol and drug consumption, infections, exercise, and aging.[68] And thus the original tests were done under false assumptions about the hormones and the subjects. The most obvious counterpoint is also the most significant: Just because the hormones of homosexual men sometimes show a particular reaction, no one has demonstrated a relationship between that reaction and sexual orientation.

The hormonal hypothesis for explaining homosexuality remains unsubstantiated. The altered sexual responses in rats mean little when considering human sexual orientation, which includes fantasy, feelings, and not always the actual sex act.[69] In any case, most homosexuals, both men and women, have perfectly normal hormones. They have normal genitalia, a normal sex drive, produce sperm and eggs, and can conceive children. Homosexual men are not men caught in an estrogen storm, nor are lesbians women who have too much testosterone. There may be prenatal hormonal changes that effect changes in the brains of some people, but as yet, science has not shown how, or even if, those invisible chemicals called hormones can change brain or body chemistry and spark the nature of sexual attraction.

WIRING

Pack five hundred people in a room and look them over. Probably the first thing you'll notice is that there are two types of people, one we call male and one we call female. You'll notice this mostly because these two types of people wear different clothes, have different hairstyles and other accessories that beg them to be separated into two distinct groups. And now take their clothes off. Those two distinct categories become even more apparent. Half of the people have breasts that stick out and the other half have penises. This is true regardless of the sexual orientation of these people, what they do for a living, or the color of their eyes. So a male versus female division makes sense, and is clear, because there isn't much overlap in these biological categories. For some scientists, this male-female division goes way beyond our hairstyles, even past our genitalia, and straight to the brain. They believe in, and spend their research time looking for, physiological differences in male and female brains.[70] This in turn has led to studies that seek the source of the "sexual brain" by comparing men and women and homosexuals and heterosexuals.

First, researchers at UCLA, led by neuroanatomist Roger Gorski, have pursued the possibility that inside the brain is the key to understanding differences between men and women in their sexual motivation. Concentrating on the rat hypothalamus, a region that regulates metabolism and is connected to some automatic behaviors such as sexuality, Gorski discovered an area at the front of the hypothalamus, or the medial preoptic area, that is twice the size in male rats as in female rats. Called the sexually dimorphic nucleus of the preoptic area (SDN-POA), it seems to be implicated in a male rat's ability to mount a female, although no one is really sure what goes on there. They do know that the size difference between the sexes develops under the influence of the presence or absence of male androgens.[71] Following this research, the same group, and teams in other parts of the world, have begun using human brains to search this particular area of the hypothalamus for clues to male-female brain differences that might account for differences in male and female sexuality. Neuroanatomists are most interested in four tiny clusters of cells called, and numbered, interstitial nuclei of the anterior hypothalamus 1–4 (INAH 1–4), first identified in the tissue of human brains by Laura Allen, a colleague of Gorski's. Various studies of these cell clusters present confusing and often contradictory results: INAH-1 has been described as larger in men than women in one study, but no difference was seen in two other studies; INAH-2 is supposedly larger in men in one study but not different from women in two other labs; INAH-3 is larger in men in one study but not in others; and no one has looked at INAH-4.[72]

In any case, the hypothalamus with its tendency to be different in males and females, and its possible connection to sexuality, most recently led neurophysiologist Simon LeVay to explore a new spin on the presumed sexual dimorphism of INAH regions. One possible explanation for homosexual orientation might be analogous to neuroanatomical male-female differences; since male homosexuals are oriented toward men, the parts of their brains that guide sexuality should be female-like.

LeVay, who also happens to be gay, based his particular hypothesis on work that showed larger regions of INAH-2 and -3 in men than in women, and the assumption that this area of the hypothalamus should have something to do with sexual motivation. Women, who are usually turned on by men, have smaller INAH-2 and -3's. Therefore LeVay figured homosexual men, who are also turned on by men, should have small INAH-2 and -3's as well. LeVay dissected the INAH regions of forty-one individuals, including nineteen homosexual men, sixteen heterosexual men, and five women. He discovered that the volume of the cells in the INAH-3 region was typically twice as small in the homosexual men than in the heterosexual men although region 2 was the same size. Also, the size of the homosexuals' INAH-3 closely matched the INAH-3 size of his six female subjects.[73] According to LeVay, then, various sizes of this area of the hypothalamus are associated not with a person's sex, but their sexual orientation.

This study has been followed by several others that report differences in areas of the brain between homosexual men and heterosexual men. For example, the UCLA group found that the anterior commissure, a bundle of tissue that sends messages from one side of the brain to another, is larger in females than in males, and even larger in homosexual males, but this result could not be reproduced in another lab.[74] Dick Swaab of the Netherlands discovered differences in the suprachiasmatic nucleus, an area tied to biological rhythms, between homosexual and heterosexual men. This particular nucleus is typically larger in women than in men and Swaab found that homosexual men had nuclei not only larger than those of heterosexual men, but also larger than those of women. The same homosexual and heterosexual male brains did not differ, the way that men and women and male and female rats do, in another hypothalamic area called the sexually dimorphic nucleus.[75] Based on the inconsistency of his findings, Swaab points out that any proposal that male homosexuals have "feminized" brains is much too simplistic, and certainly incorrect for most areas of the brain. In any case, these areas of the brain, and others that labs are currently comparing, have yet to be connected to sexuality at all.

All this scrambling for identifying the neurophysiology of the male homosexual brain has left this branch of science in a mess. Few of these studies have been replicated. And when they have been, the results are most often contradictory. The studies also suffer from major method-ological problems. Measurements of cell counts or size of a particular area of the brain may result in statistical differences among classes by sexual orientation, but they also show large amounts of overlap. For example, several homosexual men in LeVay's study had INAH-3 areas as large as heterosexual men, as did some of the heterosexual women. All of the homosexual brains used in these studies came from patients who died of AIDS, and no one really knows what happens to the hypothalamus or other areas of the brain during the ravages of the various complications of AIDS. Even more important, the sexual orientation of any of these subjects is certainly not a given. Men dying of AIDS might be confi-dently labeled as homosexual (although they might have contacted AIDS through drug use or heterosexual sex), but who knows the sexual orien-tation of the so-called heterosexual men, or of the females for that mat-ter. Also, the underlying assumption that the hypothalamus, and the INAH areas in particular, is connected to the roots of sexual motivation is shaky at best.[76] The hypothalamus and limbic system are involved in sexuality, but so are many other parts of the brain. A direct hypothala-mus-copulation conception might work for male rats, but no one has shown a clear connection between this area and sex in primates, includ-ing humans.[77] And finally, there's no way at the moment to know if these so-called differences in brain architecture are the cause of homo-sexual attraction, or perhaps the result of a lifetime of homosexual prac-tice which changed the anatomy of the brain. The science of the sexual brain has therefore left us with more questions than answers.

A more significant criticism can be leveled at the politics of these stud-ies. It seems that large sums and much research time are being spent on an attempt to understand why a small percentage of men are excited more by men than women. Although the scientists might argue that their research will help us understand the roots of sexual motivation in general, it's curious that male homosexuals have inspired such diligence. And there seems to be favoritism here; late to the lab table come the brains of homosexual women, as if their sexuality was of little interest or concern. William Byne also points out that our culture wants us to think of sexuality, and therefore brain anatomy, as only having two paths—

attraction to men or attraction to women.[78] But clearly, human sexuality is not that black and white. In fact, these studies that purport a significant difference between men and women, and homosexuals and heterosexuals, most often have a range of cell numbers that overlap and could be placed in either camp. These overlaps, rather than the constructed polarity in brain anatomy that is emphasized, echo Kinsey's notions and a more cross-cultural view of a continuum of human sexual response.

In any case, as the brain researchers themselves acknowledge, no one knows cause and effect here; for some as yet unknown biological reason, the brain might push the body toward one type of sexual behavior, or a type of consistent practice might in fact change the anatomy of the brain over a lifetime. None of these studies have declared they know the roots of homosexuality, only that male homosexual brains sometimes and in some places seem to end up wired a bit differently.

The Continuing Puzzle of Homosexuality

All this research and talk about a biological basis for homosexuality haven't led us much closer to understanding the fact of homosexuality. It might be genetic, which in turn might influence the sexual wiring of the brain, or change a person's hormonal soup. We just don't know. And yet most people, homosexual and heterosexual alike, would agree these days that there's a biological component to sexual orientation, no matter which direction it points to. If this is true, an important question must be addressed: Why is homosexuality a significant component of the sexuality of our species?

HOMOSEXUALITY IN NONHUMAN PRIMATES

If homosexuality has a biological basis, and isn't just a phenomenon born of upbringing and environment, there should be evidence of homosexual behavior in our closest relatives, the nonhuman primates. The evidence, however, is rather sparse.

It's true that male monkeys often mount other males from behind.[79] This behavior, however, appears to have little to do with achieving orgasms or any sort of sexual pleasure; penetration rarely, if ever, occurs in the natural state. It is, instead, a status ploy by the one on top. The mounter is asserting his dominance, and the mountee is submitting and accepting his lower status. This male-male interaction is most often seen in species in which the male hierarchy is of greatest importance to the social system such as macaques and baboons. Male baboons have also been known to grab each other's testicles as part of their social interactions. But this too is a sign of male-male alliance rather than sex. Male baboons do this in greeting, or to reassure each other and calm each other down during a tense moment.[80] In one species, play among young animals takes on a decidedly homosexual direction: Young male bonobos frequently suck each other's penises, which must feel good, but there also seems to be an element of fun rather than real sexuality in their intention.[81]

Homosexuality among female nonhuman primates has been clearly documented in several species—rhesus, Japanese, and stumptail macaques, the Hanuman langur, and bonobos.[82] The mechanism for female sexuality varies among species.[83] In Japanese macaques, for example, estrous females who are highly sexually motivated mount other females and rub their clitorises on the backs of their partners. They also form consortships with other females that last for days. They do this, suggests primatologist Linda Wolfe, because there often aren't enough males in the group for consortships and mating, and so, pushed by the inner drive to have sex, females make do.[84] Hanuman langurs do the same—females mount each other only when they're in estrus and rub their front pubises on the backs of other females.[85] Bonobos, on the other hand, use homosexuality just as they use heterosexuality—for social reasons. A female bonobo entering a new group usually makes her entrance easier if she approaches each resident female and presents her swelling for a genital-genital, or G-G, rub.[86] She and her partner clasp each other and bounce their swellings back and forth. Homosexuality, in this case, is a passport to group life rather than a sexual fling.

Although there is evidence of occasional dabbling in homosexuality in a few species of nonhuman primates, especially among some female monkeys and apes, the data don't strongly support a case for order-wide homosexuality. Other primates might use what we humans call homosexual behavior to relieve sexual tension, to make a social contact, or to get something from a partner. But they are never as exclusively homosexual, or as routinely homosexual, as some humans are. Only where we see extensive sexual behavior of all kinds, such as in bonobos, does homosexuality really appear as a regular feature. And so the roots of homosexuality may be glimpsed in our nonhuman primate relatives, but we seem to be the only primate where some members are exclusively homosexual as a way of life.

THE EVOLUTION OF HOMOSEXUALITY

If we accept that a certain percentage of men and women in every human society call themselves homosexual, and if we accept that since human history has been recording such information there have always been homosexual acts, the obvious question is Why? Surely, evolutionary biologists, theorists dedicated to understanding the ultimate reasons for human behavior, must have an answer for the seemingly universal appearance of homosexuality in our species.

It turns out this one has them baffled. Heterosexual behavior, with all its layers of love and lust, can be easily explained as Nature's way of making us, a sexually reproducing species, join up and pass on genes. Homosexuality, however, presents a special conundrum for biology. Here is a behavior, directly connected to reproductive success, and yet by definition, designed not to pass on genes, so how could it have possibly evolved as a mating strategy?

On the one hand, homosexuality could be considered an evolutionary aberration. Not all human behavior is adaptive, and perhaps homosexuality is just one of those patterns that show up randomly and are never passed on. Biologist Edward O. Wilson has another take. He believes that homosexuality is a genetically driven behavior that evolved because it has advantages.[87] His logic goes like this: Homosexuality is a human universal and therefore genetically based. Although many homosexuals do not pass on genes directly, they do share genes with their kin. Any helping hand a homosexual gives to his or her kin then helps pass on those more recessive, but highly altruistic, genes. According to Wilson, homosexuals carry the very rare tendency for altruism in our species, and the genes for homosexuality have been passed on along with those special and important altruism genes. He imagines that early in our ancestry, when humans lived in small kin-oriented bands, these altruistic homosexuals were integral to the health and welfare of the larger kin groups. Their altruism, and their homosexuality, are here with us today because our ancestors needed the help homosexuals gave to the band.

This is an interesting approach, and the only evolutionary one I know of. It does, however, suffer from some major loopholes. First, there's no evidence to suggest that homosexuals are more or less altruistic than heterosexuals. Maybe a homosexual uncle takes his nieces and nephews to the mall, but so does the heterosexual aunt. Taking the long view, there's absolutely no evidence that homosexuals in our ancestry had any impact on the success of bringing up others' children. Second, many homosexuals do have children. Gay men often have families, and lesbian women often have children. The idea that homosexuality has been selected for in the human species because homosexuals have spent their lives tending the genes of others is doubtful. It's an interesting possibility, but not a well-substantiated one.

We probably can't come up with a good evolutionary explanation for homosexuality because there isn't one. If homosexuality is a variant of the human sexual continuum, then we need not search for a "why" answer for its appearance. Probably all human sexuality is genetic in a sense, and probably there are genes that predispose someone to land on a particular place on the continuum. But those genes don't migrate nicely into categories of homosexual, heterosexual, and bisexual, and neither do we. The search for a biological, evolutionary, or psychological cause for homosexuality, I think, misses the point. Scientists and theo-

CHAPTER SEVEN

Sex Beyond the Twenty-first Century

The natural history of the human species is a short one. So far, we and our ancestors have lived only about 4 million years—and we really haven't changed much in those years. Our first ancestors started out as tree-living furry mammals, moving about the canopy using both arms and legs for movement and support. But at some point these early primates came down from the trees and opted for a more terrestrial life, and they stood up on hind legs to walk into uncharted areas. Along with this major shift from a quadruped to a bipedal animal, the species also lost most of its body hair. It lived on the savannas of Africa for a few million years and then eventually spread around the world, an ecologically successful species. In the short span of about 1.5 million years, although our bodies changed hardly at all, our brain grew exponentially. Today, we have the largest brain relative to body size in the animal kingdom. As a species, we seem to be happiest living in groups of varying sizes, from small families to village communities to huge metropolitan areas. Collectively, we've managed to build complex tools that require sophisticated cognitive skills and physical dexterity. We have also developed cultural artifacts that protect us from Nature. Our communication is a constant stream of chatter about the present, the past, and even the future. Oddly enough, only a small minority of our species now forages or cultivates food, and this minority is responsible for feeding everyone else. Sex is important to humans; we think about it, talk about it, and have it more often than is necessary for reproduction. Like a few other mammals, human females have only a small number of babies over a lifetime, and they produce only one infant at a time when they do conceive. The next generation is always a heavy burden—juveniles seem to hang around for years. Although many other animals, such as large carnivores, have the potential to prey on us, our own worst predator is other members of our species.

No one knows the end to this story. Most species on earth evolve into something else or eventually become extinct. Only a few have stayed the way they are for very long—sharks and bacteria come to mind. Over the history of the earth, an untold number of species have evolved, adapted, and then eventually either became extinct or adapted again. As a species, we may think we rule the earth, but we're only a very recent addition; we showed up only last night in evolutionary terms. In fact, bacteria and insects have been far more prolific and successful than we have been—they've been here much longer and have adapted to many more climate changes and ecological niches than we might ever imagine. And yet we have the audacity to assume we are the end point of evolution, doing "better" than any other species, and here to rule forever. But no one can really predict what the future will bring for *Homo sapiens*.

Humans, like all animals, continue to evolve. We've only looked the way we do now—hairless, bipedal apes with big brains—for a little over a million years. Many generations from now, if the human species is still around, there will be many changes in the ways our descendants look and behave. We might be buffered in some sense from Nature by the trappings of culture, but in reality we're no more shielded from natural selection than the fly on the screen door. Wearing clothes, constructing skyscrapers, manufacturing synthetic food are all well and good, but none of them will save us from disease, famine, or drastic climatic changes. Our only guarantee is that natural selection will affect the future of our kind. In other words, we can't exactly predict what humans will look like several hundred generations from now, we can only predict that change will come.

I often ask my students, after a semester of looking through several million years of casts and slides of human fossil bones, to make predictions about our descendants. They foresee creatures with bulbous heads and huge brains; if intelligence is the mark of our species, they suggest, large brains will be selected even more over time. Others see humans with tiny, atrophied brains that waste away as computers take over the harder mental tasks. According to my students, future members of our species will live longer, not wear glasses, never get cavities, and have babies conceived in petri dishes and gestated in artificial uteruses. They think technology will rule our lives and make life easier, which in turn will change the way our bodies evolve.

These responses always strike me as fanciful—words of wisdom from an elite group of college students well versed in science fiction. The reality of our future is actually the same as the reality of our past. Evolution occurs when populations adapt. If they don't adapt under environmental pressure they become extinct. We humans like to set ourselves apart from the ordinary rule of evolution because we wear the cloak of culture, but the knife of natural selection cuts through that costume as well. Culture, technology, civilization, high intelligence—all those human specialties—will not save us from Nature because they are essentially part of Nature as well. In the end, the decorations of culture may not make one bit of difference when the dice are rolled and Nature selects for this or that.

The rules of natural selection go like this: Those who pass on more genes are represented in future generations. That ever changing gene pool, moving down generations, is the shape and feel of the human species. And so to predict our future, we have to consider those who pass on the genes, not those who have the most intelligence, the most money, or the most technology. The face of our species is not molded by those who sit in front of computers but by those groups that have sex and make babies. Sex then is one of the keys to understanding our future. Who has sex, who conceives, who brings up children? To make predictions about our future as a species, we can't just turn to the pages of science fiction novels and decide that only those in technologically advanced countries will be the arbitrators of our future. Instead, we have to pause and take a global look at the human species. Our future selves are a combined product of the woman in the rice paddy, the man herding cattle on the African savanna, the woman in the Syrian market, and the American housewife. Those individuals are the breeding pool—the people who have sex and conceive the next generation. Taking these disparate groups into account, we can then ask, what are some forces that might influence who will have sex and pass on genes over the next several generations? Here I choose the most obvious possibilities that have the potential to affect the sexual and reproductive life of our species —AIDS, new reproductive methods, and hi-tech sexual enjoyment.

The Specter of AIDS

No book about sex written in the 1990s would be complete unless it addressed the issue of AIDS (Acquired Immune Deficiency Syndrome). Unlike other viruses that travel well in air, the virus that causes AIDS, human immunodeficiency virus (HIV), is rather fragile and is transmitted only in bodily fluids such as semen and blood.[1] HIV now comes in two types, HIV-I and the more rare HIV-II. Soon after infection the body begins to manufacture antibodies in an attempt to fight off the HIV infection. After a few months, the titer, or level of antibodies, is usually so high that an HIV-antibody test easily reacts to their presence —the person is HIV-positive. But this doesn't mean someone has AIDS, the disease. People become infected with the HIV virus and yet might show no symptoms of AIDS, the disease or complex of diseases, for several years; the average delay from infection to full-blown AIDS is ten years.[2] The HIV virus eventually destroys the immune system by entering white blood cells, or lymphocytes. It's especially destructive to T cells, the class of lymphocytes that fight off infection and disease. As the body loses its natural defenses in the T cells, infections that are normally sloughed off then take over. A person dies not from HIV infection, but from the annihilation of his or her immune system.

AIDS was first given a name in 1981, but cases of what we now know as AIDS had been sporadically reported in the literature since at least the 1970s.[3] In the early 1980s, the Centers for Disease Control in Atlanta, the CDC, noticed a sudden increase in requests from doctors in San Francisco and New York for certain drugs to fight pneumocystis carinii pneumonia (PCP). PCP is an uncommon disease that usually shows up in newborns, who have weak immune systems, or adults on immunosuppressive drugs. The CDC began an epidemiological study of the PCP patients. They soon realized that these patients had one common denominator—they were gay men. The first official death from AIDS was recorded by the CDC in 1981, but there is a strong possibility of deaths from AIDS as early as the 1950s.[4] By 1982, cases of AIDS began showing up in nongay groups including Haitians, drug users, female prostitutes, and hemophiliacs. By piecing together lifestyles and individual behavior, the team at the CDC suggested transmission by the body fluids semen and blood, sometimes with a shared needle as the vector. A group at the Pasteur Institute in France and a competing group of scientists at the National Institutes of Health in America began working on the virology of the disease. They soon identified the AIDS virus, called HIV, and eventually developed the blood test that picks up antibodies to HIV. We now know the HIV virus is transmitted when infected blood is passed into another bloodstream as a needle leaves one arm and enters another, when an HIV-positive mother gives birth and her blood is mixed with her newborn's, or when people have sex. In the recent past, before the blood supply was purged, receiving a blood transfusion was also risky because some blood supplies had been tainted by HIV-infected donated blood. AIDS is classified as a sexually transmitted disease because the HIV virus lives in semen and can be passed from one person to another vaginally or anally. This is especially true of anal sex when fragile tissues of the rectum are split and exposed, which then allows free passage of the HIV virus from the semen of the donor into the bloodstream of the recipient.

In 1990, over 10 million people were reported HIV-positive. According to the World Health Organization, the rate of HIV infection is higher in developing nations than in industrialized countries. Africa is a continent of special concern, where an estimated one in forty adults is presumably infected.[5] While only 100 people died of AIDS in America by the end of 1981, 130,000 were dead from AIDS only ten years later.

THE SOCIOLOGY OF AIDS

Some have called AIDS the plague of our generation—a disease that shows up years after transmission, causes multiple symptoms as it destroys the immune system, and has no cure or vaccination. Because it was first identified and given a name based on cases from the gay community in America, AIDS has also taken on a particular sociology. It began as a bicoastal urban phenomenon and only recently has spread throughout the country.[6] It's also true that the largest number of American AIDS cases so far are male homosexuals, but the number of new cases has recently decreased among homosexuals while increasing among heterosexuals. The most common venue of sexual transmission is still anal sex. Although male homosexuals who receive anal sex are at great risk, heterosexual couples, with the female receiving, who engage in anal sex put these women at risk as well.[7] Other subgroups such as intravenous drug users, those needing blood transfusions, and anyone who has casual sex without a condom are now at risk. Concern among medical officials is high for teenagers, a group with a growing HIV infection rate and little acceptance of their vulnerability.[8] Although Americans tend to think of AIDS as a "gay disease," the pattern of infection is different in other countries. If AIDS had been first identified in central Africa, where 40 percent of AIDS cases are women, it might have taken on the character of a disease passed between heterosexual partners and transmitted vertically by mothers to infants.[9]

Because of its well-publicized connection with the gay community in this country, AIDS is seen by some as retribution for sins against God and Nature. The idea of punishment for presumed sexual lewdness is a repeated pattern in human history. The Black Death, or bubonic plague *(Yersinia pestis)*, that came out of China and devastated Europe between 1348 and 1350 was also seen as a response to moral degeneration.[10] Syphilis was first discovered in the fifteenth century and became widespread in the sixteenth century. It too was considered a disease of the promiscuous. AIDS echoes the spread of syphilis because both are sexually transmitted, spread vertically from mothers to infants, have acute symptoms, and are judged morally by society. One historian has even suggested that the rise of Puritanism, a social movement for sexual modesty and monogamy, came about as a reaction to the ravages of syphilis.[11] Devastating diseases have also been used, or their spread ignored, when one group is invested in conquering another. For example, the Spanish brought smallpox to America which just about eliminated indigenous groups on the continents in the short span of fifty years and helped the Spanish to conquer the "New World."

Certainly there are those who believe that gays and drug users with AIDS have gotten just retribution and resent any efforts to stop the spread among those groups. Another group believes AIDS is reasonably contained within the high-risk groups, and education and prevention efforts should be focused on those groups.[12] More sober voices know that HIV infection has since entered the mainstream population and presents a major health risk for anyone who has sex at all. Without a vaccine, the only way to curb the spread of AIDS is through altered human behavior during sex and drug use.

This disease, often sexually transmitted, has the potential to change the behavior of millions of people all over the world. Of interest to the subject of this book is the possibility that concern over HIV infection might alter forever the way our species has sex. The possibility is there, and if we were acting in our best health interests, fear of contagion would change our sexual behavior, at least until the virus is eliminated. Has there been a change in human sexuality since the advent of AIDS?

THE EVOLUTIONARY EFFECT OF AIDS

All viruses are subject to natural selection. We're encouraged to get a flu shot each fall because the mix of viruses spreading across the globe changes as old viruses mutate into new ones and the old vaccines are rendered useless. Old viruses are wiped out by natural selection, or they become new ones through mutation. We're in a continual arms race with the flu; the flu viruses adapt to our vaccines, and so our vaccines adapt again to them. The HIV virus has probably been around forever, but it was once held in check by natural selection. It found vigor when changes in human behavior opened up a vast arena for new hosts. If there was no intravenous drug use, no anal sex, no sex with multiple partners, and less easy travel across the globe, HIV would still be one virus among many instead of a virus that was able to cross a critical threshold of transmission and become a pandemic.[13]

AIDS is interesting from an evolutionary view because it seems to be changing the way the human species views sex, that is, the way we talk about sex. Less well known is how the possibility of HIV infection has changed, or is changing, how we go about having sex. Clearly, some people are more at risk than others—intravenous drug users, infants of HIV-positive mothers, promiscuous male homosexuals, heterosexual women who sleep with HIV-infected men. And yet, the specter of this disease has the potential to change the way people view the act of sex, sex with multiple partners, and eroticism. The most dramatic effect on sexual practice so far has been documented in the gay community, where casual sex is reported to be slightly less frequent since AIDS.[14] More important, even when the number of sexual partners remains the same among gay men, most of that contact is now safe sex with a condom. The effect of AIDS on heterosexuals is more difficult to assess. *The Janus Report* claims that although people say they are more cautious about sex, that is by screening partners or using condoms, sexual activity has actually increased in recent years.[15] A study of the sex life of over twenty thousand people in France shows that those with riskier sexual behavior, that is homosexuals or heterosexuals with multiple partners, say they are more likely to use a condom.[16] This survey suggests that sexually active individuals are more aware of their risk level and often, but certainly not always, act accordingly. It also looks like education has had an impact on many young adults who have been bombarded with the message. Younger adults state that they use condoms much more frequently than older adults, although condom use in general is much lower than it ought to be given the overall risk.[17] In fact, today's teenagers, the first generation to have AIDS as part of their sex lives from the moment they lose their virginity, appear to be magnificently oblivious to the possibility of infection. Even among my peers, older and supposedly aware adults, I hear stories of casual unprotected sex with unlikely people.

The problem is that sex, by definition, is an impulsive behavior. A person is pushed by their physiological and psychological makeup to seek out sex or respond to the advances of another. The only absolute protection against AIDS is abstinence, and I doubt that many people will adopt that policy. We certainly won't stop AIDS by advocating celibacy, because humans are sexual animals designed by Nature to have a healthy sex life. We might be able to curb AIDS by advocating monogamy, but as I've shown in previous chapters, we don't seem to be naturally monogamous creatures and I don't think it is within our power, as a culture, to change the natural sexuality of a species.

But perhaps culture can be of use in protecting against AIDS. A secondary protection against HIV infection is afforded by wearing condoms. Many people refuse to use them because a condom changes the feel of sexual intercourse. Maybe our only hope as a species to curb the spread of AIDS is to use the mantle of culture to press for sex with a condom as the normal way people of all cultures have sex. If societal norms can force people to wear clothes even when they don't need them, and make mandatory elaborate celebrations even when they have no practical effect, it can surely advocate "dressed-up" sex as the normal ritual of sexual interaction.

THE FUTURE GENE POOL

Will AIDS have a real evolutionary effect on the population? Dying of AIDS obviously terminates the passing of genes to the next generation.

If HIV infection increases at the rate predicted by demographers, a significant number of people will die from AIDS in the near future.[18] Some of these people will die before they reproduce. AIDS then has the potential to change gene frequencies in the next generation. But how will the gene pool change? Those who believe AIDS was sent here as punishment for unnatural sexual behavior have suggested AIDS will wipe out most gay males. If the research on homosexuality is right, and there is a genetic predisposition for homosexuality, then these extremists might be correct from a biological perspective—in a biological sense, the loss of these "gay genes" would change the gene pool. But homosexual men are less likely to pass on their genes anyway because they have fewer opportunities for having children. More important, the preliminary research shows that whatever DNA might predispose a man to be gay is passed down through women anyway, so the point is moot. If male homosexuality is genetic and passed on though women, AIDS will have no effect wiping out these genes. In any case, this line of reasoning is extremely narrow in scope given the widespread, that is heterosexual, nature of the disease. AIDS is certainly not restricted to the gay population across the globe and therefore not destined to alter the gene pool in that way. As the brochures at my local AIDS Work office point out, it's not who you are but what you do that puts you at risk for contracting HIV.

But there's another way HIV might have a significant effect on which genes are passed on. One of the odd facts of HIV is that some people repeatedly exposed to the virus never become infected, while others, exposed only once, are infected. In addition, the lag time from infection with HIV and outward symptoms of AIDS varies widely; some people become ill rather quickly while others show no signs ten years later. There must be, therefore, different immuno-constitutions among our species. Although AIDS strikes regardless of race, religion, ethnicity, or sexual orientation, it does seem, at times, immune to some sort of physiological barrier that no one has yet discovered. In the future, perhaps, researchers may discover a genetic predisposition, some special character of the immune system, that protects certain people and makes others more vulnerable. If so, people with immune systems more resistant to

HIV, those who contract but are able to fight HIV, might pass on more genes than those who are quickly felled by the disease. We could see natural selection acting against those with less able immune systems, and selecting for those who can fight this pandemic.

In reality, it's hard to predict what effect AIDS might have on the gene pool of future generations at this stage because the social and environmental factors are completely unknowable. In the short span of fifteen years since the disease was first identified, some people profess to have changed their sexual habits. It may be that safe sex, which would also curb the rate of pregnancy, since condoms also serve as contraceptives, will become the "normal" kind of sex for everybody. Or there might be a vaccine developed in the next many years that will make AIDS, like measles, a vanishing disease, in which case the long-term effect of AIDS on the sex life of our species will be minimal. For now, all we can say is that we stand here in the midst of a plague, and each one of us is a potential vector, host, or end point.

Reproduction in the Future

Just about every baby in the world is made like this: A woman and a man have sex, the woman becomes pregnant, and roughly nine months later the baby is born. Medical technology over the last two decades has changed all that. Suddenly, there are scores of infants conceived and born in much more creative ways. We now have frozen sperm, frozen embryos, surrogate mothers, egg donors, sperm donors, *in vitro* fertilization, artificial insemination, and post-menopausal moms. Each week a new technique comes that makes us think reproduction has much more to do with technology than old-fashioned sex. This development is especially acute for our species because unlike most other animals, humans often have sex with no reproductive consequences. Sex was, until recently, something we did for pleasure, as well as a way to make babies. With new technology, it looks like there's a potential to decouple the two, sex and reproduction, completely and forever. Are there possible consequences to our sex lives?

THE BABY RECIPE

Reproductive technologies weren't invented just for fun as a new way to make a baby. They were developed to help infertile couples make biological children.[19] Infertility strikes about one out of every five couples in Europe and America. No one knows if this is the result of a sudden increase in infertility, or if people are just more vocal about their infertility and more often seek medical intervention than they did in the past. There does seem to be an increase in the number of women delaying childbirth in industrialized nations, which decreases fertility and might account for some of the higher numbers. Also, in the last forty years the average sperm count has decreased rapidly, presumably as an effect of toxins in industrialized environments. The causes of infertility, when diagnosed, lie equally with the man or the woman, or both parties have a problem that contributes to the overall picture.[20] Medical science, which knows so little about conception and gestation in the first place, is at a loss to explain infertility in about 15 percent of the cases. Clearly, some people are more or less fertile than others, just as some people have better or worse hearts, kidneys, livers, and stomachs. The lesson here is that fertility, the ability to have sex and reproduce, is not a given; as in all things biological, there is variation.

The medical community labels a couple as infertile if they're unable to produce a live infant after one year of unprotected sex. Sometimes, this is just a matter of odds. The average woman has only a 15 percent chance of conceiving each cycle, but many cycles are nonovulatory, and certainly no one always has sex at the best moment during ovulation. These odds fly out the window for those women who are infertile, because they don't have reproductive systems that function very well in the first place. One study has shown that seeking medical intervention for infertility is not necessarily useful. In one group diagnosed with infertility problems, half chose to pursue medical intervention and half walked away from modern medicine. The rate of subsequent pregnancies was the same in both groups.[21] And the idea that adoption will somehow correct infertility is simply an old wives' tale; for those who adopt and those who don't adopt, the pregnancy rate is the same, about 5 percent.

Psychologically, infertility is devastating for those who want a family.[22] Childless couples are surrounded by other couples with children moving blithely through this lifecycle stage. Reproductive physiologists have tried to ease the pain of infertility by inventing various procedures developed to improve the chances of conception. A woman with endometriosis, a condition in which uterine tissue grows outside the uterus, for example, can have her tubes unblocked; women who ovulate sporadically or not at all can take drugs to regulate their cycles; men with low sperm counts can take drugs or remove testicular obstructions to improve their sperm count and motility; couples can also conceive outside the womb or opt for using the eggs and sperm of others, or hire a female gestating body for the care of their externally conceived embryo. Medicine has developed these options as correctives to reproductive systems, much as they have developed fake legs for amputees. These substitutes for a decent set of reproductive organs help, but not all that much. Contrary to stories in the popular media, all these reproductive technologies just aren't very successful. *In vitro* fertilization has a very low rate of success; infants conceived on fertility drugs have a higher rate of miscarriage than infants not exposed to that chemistry, and in general, people with fertility problems have a higher rate of miscarriage; using surrogates opens the door to endless, and expensive, legal battles. In other words, the technology is not exactly a realistic fix in this situation.

On the other hand, these new methods to help Nature along could be used by the population at large. The fertile, rather than the infertile, might avail themselves of some of these techniques to change the usual process of conception, pregnancy, and birth.[23] But still, these methods are uncomfortable, perhaps dangerous in the long run, not very successful, and not as much fun as sex. Until medical science replaces reproduction with technologies that are low-cost and more successful, most people will pass on genes the usual way.

WHAT IF WE DIDN'T NEED SEX?

It takes a lot of energy to have sex. Just ask a sixteen-year-old boy who spends much of his time thinking about sex, fantasizing about a willing partner, and if he's lucky, actually having sex. Gloria Steinem recently wrote that one of the joys of being post-menopausal was that she no longer felt that all-compelling drive to look for sex, and she had so much more energy for other fun things.[24] Although no one in good health is ever sexless, regardless of age, there is usually a drop in sex drive with age for both men and women.[25] Think of a pie chart of each person's daily energy. It takes so much energy to move about, work, prepare food, clean oneself, and have sex. Wipe out the sex part and the energy can be used for other things. Natural selection might swiftly move to favor those who passed on genes efficiently in a new-age sort of way if that freed-up energy was used in other advantageous ways during what should be the best reproductive years. If a person could pass on her or his genes without all the mess of sex and fuss of pregnancy and childrearing, might not natural selection favor that kind of reproduction?

The potential for a change in human sexuality might be possible if we really had techniques for reproduction outside of the body. The only problem is that what we have so far is rather primitive; no one is saving any energy here. Let's start with *in vitro* fertilization, the mixing of a sperm and egg in a glass dish. Maybe conception occurs in glass, but that few-celled zygote is quickly transferred into a human body. Time and energy saved—none. The man still had to masturbate and ejaculate and the woman still produced that egg. More to the point, *in vitro* is notoriously unsuccessful. The technique is relatively primitive and only an infertile couple with lots of money for this expensive procedure would even bother. Surrogacy, on the other hand, works pretty well once you find a candidate. The man masturbates and ejaculates and his sperm is used to impregnate a woman. But the surrogate still has to gestate and deliver. No reproductive energy has been saved—it's merely been transferred to another woman. So far, no one has figured out how to conduct pregnancy outside of a female body. In the end, what looks like space-age reproduction is really only the mixing of eggs and sperm in different media. Gestation and childrearing remain the same—a high energy drain for a long-lived species.

More important to any consideration of the future of our species, these technical ways for conception are available to only the tiniest minority of people—that's why they're news. Most of these techniques were developed on, and for, people who are infertile, people who have tried forever to make babies the normal way. These couples turn to technology as a last resort, not a first choice. And when they do, the cost of such procedures is enormous. Notice how no one seems to invent new reproductive technologies in Third World countries where populations are burgeoning. Technical conceptions are for an elite, rich, medically sophisticated population. So these new-age babies are possibly only for the very few who can afford the cost of building families this way.

It's not likely that test tube babies will be the norm anytime soon. Until all the women of our species get their tubes clipped and all the men have vasectomies, and we figure out a way to gestate babies without a uterus, sex will be around because it will, for a very long time, be the only reasonable way to reproduce.

The Naked Ape in Sexual Cyberspace

It's easy these days to have sex with a stranger, especially if you have a computer. Boot up the computer, dial through the modem, and join the network of millions of linked computer users around the world. Or slip a disc into the computer's CD player and call up a lively young lady who is designed to share your fantasies. Connecting with a sexual partner has always been a major feature of any sexually reproducing species. But we're the first generation to take the actual animal out of the connection process. We've invented something much more efficient. We don't have to seek a partner, evaluate him or her, or even really interact. Eroticism can be found on the airwaves.

ELECTRONIC SEX

The internet, or "net," is a system that interconnects any computer with a telephone link to other computers.[26] For me, this link passes from my home computer to the larger university system. This allows me to travel electronically from the university to anyone else who is hooked up the same way. My favorite use of the net is electronic mail, or E-mail. I log on to the computer, join the university system, and open my electronic mailbox. When I log on today, I have three messages. One is from a friend in Michigan and another from a former student. The third is from a colleague in Scotland. We're passing messages back and forth about accommodations at an upcoming conference. After I send them messages back, I write to a colleague in Japan about the possibility of visiting him next year. I could just as easily call all these people, or write them letters, but the value of E-mail is that for me, as an employee of a university, this format is free. It costs me nothing to talk to Japan or Scotland and I know my friends will get the message within minutes. With the time difference, they might not see it until later, but they will respond, mostly because E-mail is so immediate and can't be shuffled under a stack of papers on a desk; it's always in your face if you log on. E-mail is only one way to use the net. There are also interactive user groups where people with the same interests interact. I belong to a group called Primate-Talk, where short notes about monkeys and apes are passed among six hundred primatologists. It's a good way to find out what people are doing without going to a conference, or to get advice about a research project. I could belong to any one of over forty thousand groups—name a hobby or interest and there's an internet group for it.

Bill Gates, CEO of Microsoft, says the internet is just another way for people to talk to each other. And he's right. It may be a highly technical invention, but it's just another way for people to chat about this and that. First we had face-to-face conversation. Then came the telegraph, the telephone, television, facsimile machines, and now, internet. In all these cases, information about one another is whizzing over the airwaves to another person. To me, philosophical anthropologist and student of human behavior, the internet is just another way primates have invented so that they can do what they do best—be social. In a sense, our most sophisticated technical achievement reveals our most primitive social instinct—to know what everyone else is doing.

It's no surprise, therefore, that sex has found a place on the net. For humans, it seems, sex flows closely on the heels of sociality. Form a group, and some members will discover each other and start having sex. Invent television, which is basically a means to show some primates what other primates are doing, and some of them will be shown having sex. Wire up a community for telephone access where people can talk with each other, and a certain enterprising number will use the wires to have sex. All the same ingredients for sex are there on the internet too: available partners; a way to communicate; good visuals; and the added bonus of anonymity if you want it. The internet now boasts several sex channels, which include groups such as ThrobNet, SwingNet, StudNet, etc.[27] What you get from these groups are pictures of people who probably don't really exist, or photographs of real people who were scanned into the computer's database and used over and over, like a moving picture. You can have an ongoing chat with a computer sex kitten at one of these stations, a woman created by a computer programmer, who then responds in writing to your provocative messages. More realistically, you can exchange fantasies with real people on some networks who are interested in exchanging verbal eroticism with an anonymous partner. Some people eventually make a date and meet up, but most of this futuristic sexuality is confined to bits and bytes, certainly the safest kind of sex of all, because there is no danger of any exchange of bodily fluids—just an exchange of words and pictures.

MORE THAN JUST CHAT

Many years ago I visited the offices of a friend who had just been networked with computers. "Come here," he said. "I want to show you something on my new computer." With the push of a few buttons, a black and white screen filled with a very fuzzy picture of a woman in underwear. The possible menu items over her head offered options like "undress," "spread legs," "kiss." We opted for a few of those and our friendly computer doll complied, in a jerky cartoon kind of way. I don't think my friend found this particularly appealing, but he did find it amusing. His computer was barely out of the box when his highly educated, intellectually sophisticated colleagues installed pornography on it. This route to eroticism, not an exchange on the net but programs designed for the individual user at home, has gotten much more sophisticated in recent years with the inclusion of CD-ROM capabilities in some computers. Although the CD is supposedly there to replace floppy disks and store large amounts of data, it also works well as an interactive picture and text format. Most often, this technology is used as a new way to experience books, encyclopedias, and other high-information venues. For example, my computer has a CD player, and I've "read" a book this way—a process of looking at pictures, reading text, listening to music or voices, and then selecting for further detailed information on the side. CD-ROM adapts well to sex because the computer gives you nice color visuals, a voice that talks to you, and it's a program that can be experienced all alone. You punch in the command keys and the person on the screen responds to your wishes. It's like making a motion picture figure move at will, but this time the figure, typically she, moves solely for your sexual pleasure in what is basically computer pornography.[28]

The new kind of eroticism, both on CD programs and through the internet, is available to anyone with a computer and a modem. While this might seem like a global invitation to an orgy, or a ticket to private nightly sex with a computer-generated love slave, the reality of computer sex is that only a small percentage of people really have this option open to them. Given individual poverty in Third World countries, I don't see the majority of people owning a computer anytime soon when they can't even afford shoes or food. Computer sex, at this point, is an outlet for only a small minority of our species. It's fun, it's different, but it's essentially a toy. Any impact on our species, from an evolutionary view, seems unlikely. But as I write these words on a computer, and think about all that E-mail waiting for me at the university system, I wonder if I might be wrong here. Perhaps this is a better prediction: Once we find a way to make computers reach out and touch, and once we figure out how they could be pregnant instead of women, computer sex as the accepted norm is right around the next byte.

VIRTUAL SEX

Let's face it, having sex can be a real chore. First, you have to find a partner—a person who wants to have sex with you as much as you want to have sex with them. And then you have to worry about doing things to make your partner happy, and of course there's the mess of undressing, and dressing, and so on. How much better to have sex alone. Sure, masturbation can, to a degree, fill in the gap, but there's now a process that can give you all the mental and physical stimulation of a real partner without the complications of his or her presence. Virtual reality (VR) encompasses several kinds of computer technologies that give a person the sensation that something is happening when it really isn't.[29] In the designs currently available, the operator dons a large helmet-like device that covers the eyes and ears; visual and auditory stimulation are piped in. In addition, an electronic armature is attached to various parts of the body, such as a wired glove, that sends and receives electronic signals relative to the movement of the wearer. The brain is immersed in a virtual environment that is completely created by the equipment but can be manipulated by the individual. One of my friends tells about a session with virtual reality that sent him on a flight through the Grand Canyon as a soaring dinosaur. He could move his head down and see the canyon floor below, or move it right and see other flying dinosaurs. It felt, he says, as if he were really flying.

Imagine a headdress and body sensors that could give us the experience of sex. A few electrodes on the genitals and skin surface and away we go. You could have sex with any Hollywood star and he or she would never know. Nor would your spouse. We're not just talking about visual images here, but all the physical sensations of touching someone and being touched. It may sound bizarre, but it's surely just a short step from virtual flying dinosaurs to virtual sex. Once again, however, this is going to be a playland for the very rich. Perhaps I'm wrong, or shortsighted, but I don't see virtual reality sex helmets in each of our bedrooms in the near future. I also don't envision this kind of singular sex becoming the preeminent mode of sex because sex, the act, is a need engraved onto our psyches as a social event, a need to interact with another person. If we, as a species, were content with solo sex, masturbation would long ago have risen as our primary mode of sexual release. As I mentioned, for the moment evolution won't really allow us to subsist sexually by strapping on a computer helmet and doing it alone—that is, unless we find a more efficient way to combine eggs and sperm, and figure out a way to bypass gestation. In that case, natural selection might well opt for virtual sex, virtual conception, and virtual evolution.

What's Love Got to Do with It?

Well, nothing really. No one has to love his or her partner to have sex with them or conceive an infant with them. Over millions of generations, sexually reproducing species have found mates, copulated, conceived, and produced offspring, and love wasn't really the point. The same thing can be found in our own highly conscious, supposedly intelligent species. Babies are born all the time to parents who weren't in love when they had sex, and probably didn't share a single thought of *amour* when their genes entwined and were passed on. In that sense, sex is disconnected from love for us and for all other reproducing animals. And yet I know that everyone reading this book has felt love for someone else, and perhaps has had sex with that person. They might even have had sex with someone they love, made a baby, and seen the baby grow up and make babies of his or her own. Sex, in that sense, can be a significant part of a loving relationship. It isn't a necessary part, but it does add to the richness of life. And so love is not necessarily a part of sex, but often, sex is part of love.

In this book, I have been interested in exploring humans as sexual animals. I have demonstrated that our genes, our particular evolutionary background, our specific body shape and chemistry, and our individual psyche all determine how each one of us goes about his or her sexual business. I should also add that sex is more complex than even these many pages outline. This book only adds a few pieces to the puzzle, and at times I hope shattered certain pieces that, until very recently, fit pretty comfortably. This is how science works: There are no definitive answers here or anywhere else; there is only the process of asking questions, seeking answers, and then revising our thinking. If we're lucky, it makes some sort of sense. Perhaps many years from now, after hundreds of personal surveys and miles of genetic maps, we might know what really makes the human sexual animal tick. And just when we find out some answers, I'm sure the questions will change.

When most people travel, they look for, and are excited by, the differences in other cultures. The clothes, the food, the jewelry, and the houses are often strikingly different. And if you've ever sat through an evening of travel slides by your dear friends, you know that photographs of travels always highlight these often outlandish differences between human groups. Look at the Balinese woman carrying fruit on her head. Check out that African child with the wooden spear. But when I travel, I'm most struck by what people have in common. I always enjoy gossiping with women about their neighbors in Indonesia, talking with a Maasai about the lines of his family, and staring along with everyone else at new babies in a Peruvian market. For me, these moments speak to our universal nature, our human nature, even our more basic primate nature. And when I think about this universal nature, I can easily extend it to matters sexual. I suspect having sex with an aboriginal tribesman from New Guinea wouldn't be much different from having sex with an American tax attorney. Different clothes maybe, but the act would be pretty much the same when you get down to it. Everyone, every human of our species, wants sex, affection, touching, orgasms. We're individuals, and yet linked to a common universal whole because we're all the same species, all compelled to pass on genes, all looking for love in some form or another—sometimes even in the right places.

And yet we understand so little of ourselves as organisms. At most we know that we're one biological species that came from Africa 4 million years ago and developed a culture that allowed us to spread across the globe. But we seem so confused about some of our most important parts. No one knows for sure why we're intelligent, why we became bipedal, or where we got this penchant for tools. Like all other sexually reproducing species, we need sexuality to pass on genes, but we seem at a loss to explain the place of that sexuality in our lives, especially when we use it for means other than procreation. I believe our confusion is often the result of a blind rush to see ourselves as something better, higher, greater, and more worthy than other animals. And in doing so, we have ignored our most basic primate, that is animal, nature. We can watch TV, we can exchange bits of information on the internet, we can put on lipstick and play tennis, but we're still those primates that came out of the trees not so long ago in search of food. It's time, at this point in our evolution, to stop and look inward and backward, and so much more carefully, to figure out this creature that looked across the African savanna and headed north.

This is truly the crux of our existence, because in the end, if we aren't careful, we run the risk of thinking we're better than Nature, more manipulative than natural selection, and more powerful than evolution—and there's certainly no future in that.

Bibliography

Abplanalp, J. M., R. M. Rose, A. Donnelley and F. Livingston-Vaughn (1979). Psychoendocrinology of the menstrual cycle: II. The relationship between enjoyment of activities, mood, and reproductive hormones. *Psychosomat. Med.* **41**: 605–615.

Adam, B. D. (1986). Age, structure, and sexuality; Reflections on the anthropological evidence on homosexual relations. *Anthropology and Homosexual Behavior*. Ed. E. Blackwood. New York, Haworth Press. 19–33.

Adams, D. B., A. R. Gould and A. D. Burt (1978). Rise in female-initiated sexual activity at ovulation and its suppression by oral contraceptives. *New Eng. J. Med.* **299**: 1145–1150.

Adkins-Regan, E. (1988). Sex hormones and sexual orientation in animals. *Psychobiol.* **16**: 335–347.

Akers, J. and C. Conaway (1979). Female homosexual behavior in *Macaca mulatta*. *Arch. Sex. Behav.* **8**: 63–80.

Alcock, J. (1980). Beyond the sociobiology of sexuality: Predictive hypotheses. *Brain and Behav. Sci.* **3**: 181–182.

Alcock, J. (1987). Ardent adaptationists. *Nat. Hist.* **96**: 4.

Alexander, R. D. (1990). How did humans evolve? *U. Mich. Special. Pub.* **1**: 1–38.

Alexander, R. D. and K. M. Noonan (1979). Concealment of ovulation, parental care, and human social evolution. *Evolutionary Biology and Human Social Behavior*. Ed. N. A. Chagnon and W. G. Irons. North Scituate, Duxbury Press. 436–453.

Allen, L. S. and R. A. Gorski (1992). Sexual orientation and the size of the anterior commissure in the human brain. *Proc. Nat. Acad. Sci.* **89**: 7199–7202.

Allen, L. S., M. Hines, J. E. Shryne and R. A. Gorski (1989). Two sexually dimorphic cell groups in the human brain. *J. Neurosci.* **9:** 497–506.

Altman, D. (1986). *AIDS and the New Puritanism.* London, Pluto Press.

Amann, R. P. and S. S. Howards (1980). Daily spermatozoal production and epididymal spermatozoal reserves of the human male. *J. Urology* **124:** 211–215.

Andelman, S. J. (1987). Evolution of concealed ovulation in vervet monkeys *(Cercopithecus aethiops). Amer. Nat.* **129:** 785–799.

Anderson, J. L. (1988). Breasts, hips and buttocks revisited. *Ethol. and Sociobiol.* **9:** 319–324.

Asso, D. (1983). *The Real Menstrual Cycle.* New York, John Wiley and Sons.

Austin, C. R. (1975). Sperm fertility, viability, and persistence in the female tract. *J. Repro. Fert. Supp.* **22:** 75–89.

Avers, C. J. (1974). *Biology of Sex.* New York, John Wiley and Sons.

Bailey, J. M. and R. C. Pillard (1991). A genetic study of male sexual orientation. *Arch. Gen. Psych.* **48:** 1089–1096.

Bailey, J. M., R. C. Pillard, M. C. Neale and Y. Agyei (1993). Heritable factors influencing sexual orientation in women. *Arch. Gen. Psych.* **50:** 217–223.

Bailey, J. M., L. Willerman and C. Parks (1991). A test of the maternal stress theory of human male homosexuality. *Arch. Sex. Behav.* **20:** 277–293.

Baker, R. R. and M. A. Bellis (1988). "Kamikaze" sperm in mammals? *Animal Behav.* **36:** 937–980.

Baker, R. R. and M. A. Bellis (1989a). Elaboration of the kamikaze sperm hypothesis; A reply to Harcourt. *Animal Behav.* **37:** 865–867.

Baker, R. R. and M. A. Bellis (1989b). Number of sperm in human ejaculates varies in accordance with sperm competition theory. *Animal Behav.* **37:** 867–869.

Baker, R. R. and M. A. Bellis (1993a). Human sperm competition: Ejaculate manipulation by females and a function for the female orgasm. *Animal Behav.* **46:** 887–909.

Baker, R. R. and M. A. Bellis (1993b). Human sperm competition: Ejaculate adjustment by males and the function of masturbation. *Animal Behav.* **46:** 861–885.

Bancroft, J. (1980). Human sexual behavior. *Human Sexuality.* Ed. C. R. Austin and R. V. Short. Cambridge, Cambridge University Press.

Bancroft, J. (1983). *Human Sexuality and Its Problems.* London, Churchill Livingstone.

Bancroft, J. (1987). Hormones, sexuality and fertility in women. *J. Zool. Lond.* **213:** 445–454.

Bancroft, J., D. Sanders, D. Davidson and P. Warner (1983). Mood, sexuality, hormones and the menstrual cycle. III. Sexuality and the role of androgens. *Psychosomat. Med.* **45:** 509–516.

Baringa, M. (1991). Is homosexuality biological? *Science* **253:** 956–958.

Barret-Ducrocq, F. (1991). *Love in the Time of Victoria.* London, Penguin.

Bayer, R. (1987). *Homosexuality and American Psychiatry: The Politics of Diagnosis.* Princeton, Princeton University Press.

Beach, F. (1976). Sexual attractivity, proceptivity, and receptivity in female mammals. *Arch. Sex. Behav.* **14:** 529–537.

Bedford, J. M. (1977). Evolution of the scrotum; The epididymis as primate mover?

Reproduction and Evolution. Ed. J. G. Calaby and C. H. Tyndale-Boscoe. Canberra, Australian Academy of Sciences. 171–182.

Bell, A. P., M. S. Weinberg and S. K. Hammsersmith (1981). *Sexual Preference: Its Development in Men and Women.* Bloomington, Indiana University Press.

Bem, S. (1993). *The Lenses of Gender.* New Haven, Yale University Press.

Benedek, T. (1952). *Psychosexual Functions in Women: Studies in Psychosomatic Medicine.* New York, Ronald Press Company.

Benedek, T. and B. B. Rubenstein (1939a). The correlation of ovarian activity and psychodynamic processes. The ovulative phase. *Psychosomat. Med.* **1**: 245–270.

Benedek, T. and B. B. Rubenstein (1939b). The correlation of ovarian activity and psychodynamic processes. II. The menstrual phase. *Psychosomat. Med.* **1**: 461–485.

Benshoff, L. and R. Thornhill (1979). The evolution of monogamy and concealed ovulation in humans. *J. Soc. and Biol. Struct.* **2**: 95–106.

Bermant, G. and J. M. Davidson (1974). *Biological Basis of Sexual Behavior.* New York, Harper and Row.

Bernstein, H., F. A. Hopf and R. E. Michod (1988). Is meiotic recombination an adaptation for repairing DNA, producing genetic variation, or both? *The Evolution of Sex.* Ed. R. E. Michod and B. R. Levin. Sunderland, Sinauer Assoc. 139–160.

Berscheid, E., K. Dion, E. H. Walster and G. W. Walster (1971). Physical attractiveness and dating choice: A test of the matching hypothesis. *J. Exper. Soc. Psychol.* **7**: 173–189.

Berscheid, E. and E. H. Walster (1974). A little bit about love. *Foundations of Interpersonal Attraction.* Ed. T. L. Huston. New York, Academic Press. 355–381.

Betzig, L. (1989). Causes of conjugal dissolution: A cross-cultural study. *Curr. Anthropol.* **30**: 654–674.

Bieber, I., H. J. Dain, P. R. Dince, M. G. Drellich, H. G. Grand, R. H. Gundlach, M. W. Kremer, A. H. Rifkin, C. B. Wilbur and T. B. Bieber (1962). *Homosexuality: A Psychoanalytic Study.* New York, Basic Books.

Blaffer Hrdy, S. (1979). The evolution of human sexuality: The latest word and the last. *Quart. Rev. Biol.* **54**: 309–314.

Blaffer Hrdy, S. (1981). *The Woman That Never Evolved.* Cambridge, Harvard University Press.

Blaffer Hrdy, S. (1983). Heat Loss. *Science '83* **October:** 73–78.

Blaffer Hrdy, S. and W. Bennett (1981). Lucy's husband; What did he stand for? *Harvard Mag.* **July-August:** 7–10/46.

Blaffer Hrdy, S. and P. L. Whitten (1987). Patterning of sexual activity. *Primate Societies.* Ed. B. B. Smuts, D. L. Cheney, R. M. Seyfarth, R. W. Wrangham and T. T. Strusaker. Chicago, University of Chicago Press. 370–384.

Bohlen, J. G., J. P. Held and M. O. Sanderson (1980). The male orgasm: Pelvic contractions measured by anal probe. *Arch. Sex. Behav.* **9**: 503–521.

Bohlen, J. G., J. P. Held, M. O. Sanderson and A. Ahlgren (1982a). The female orgasm: Pelvic contractions. *Arch. Sex. Behav.* **11**: 367–386.

Bohlen, J. G., J. P. Held, M. O. Sanderson and C. M. Boyer (1982b). Development of a woman's multiple orgasm pattern: A research case report. *J. Sex Res.* **18**: 130–145.

Boswell, J. (1990). Social history: disease and homosexuality. *AIDS and Sex: An*

Integrated Biomedical and Biobehavioral Approach. Ed. B. Voller, J. M. Reinisch and M. Gottlieb. Oxford, Oxford University Press. 171–182.

Bromberg, J. (1988). *Fasting Girls; The History of Anorexia Nervosa.* Cambridge, Harvard University Press.

Broude, G. J. and S. J. Greene (1976). Cross-cultural codes on twenty sexual attitudes and practices. *Ethnology* **60:** 409–429.

Burley, N. (1979). The evolution of concealed ovulation. *Amer. Nat.* **114:** 835–858.

Burt, A. (1992). "Concealed ovulation" and sexual signals in primates. *Folia Primatol* **58:** 1–6.

Burton, F. B. (1971). Sexual climax in female *Macaca mulatta. Proc. Int. Cong. Primatol.* **3:** 180–191.

Buss, D. M. (1985). Human mate selection. *Amer. Sci.* **73:** 47–51.

Buss, D. M. (1986). Sex differences in human mate choice criteria: An evolutionary perspective. *Sociobiology and Psychology: Ideas, Issues, and Applications.* Ed. C. Crawford, M. Smith and D. Krebs. Hillsdale, Lawrence Erlbaum Associates. 335–351.

Buss, D. M. (1989). Sex differences in human mate preferences: Evolutionary hypotheses tested in 37 cultures. *Behav. Brain. Sci.* **12:** 12–49.

Buss, D. M. (1994). *The Evolution of Desire.* New York, Basic Books.

Buss, D. M. and D. P. Schmitt (1993). Sexual strategies theory: A contextual evolutionary analysis of human mating. *Psychol. Review* **100:** 204–232.

Butler, P. M. and W. H. Peeler (1979). Models of female orgasm. *Arch. Sex. Behav.* **8:** 405–423.

Byne, W. (1994). The biological evidence challenged. *Sci. Amer.* **270:** 50–55.

Byne, W. (in press). Science and belief; Psychobiological research on sexual orientation. *Biology and Sexual Orientation.* Ed. J. De Cecco. New York, Haworth Press.

Byne, W. and B. Parsons (1993). Human sexual orientation: The biological theories reappraised. *Arch. Gen. Psych.* **50:** 228–239.

Callender, C. and L. M. Kochems (1986). Men and not-men: male gender-mixing statuses and homosexuality. *Anthropology and Homosexual Behavior.* Ed. E. Blackwood. New York, Haworth Press. 165–178.

Cant, J. G. H. (1981). Hypothesis for the evolution of human breasts and buttocks. *Amer. Nat.* **117:** 199–204.

Caro, T. M. (1987). Human breasts: Unsupported hypotheses reviewed. *Human Evol.* **2:** 271–282.

Caro, T. M. and D. W. Sellen (1990). The reproductive advantages of fat in women. *Ethol. and Sociobiol.* **11:** 51–66.

Carpenter, C. R. (1942). Sexual behavior of free ranging rhesus monkeys *(Macaca mulatta). J. Comp. Psychol.* **33:** 113–142.

Carrier, J. M. (1980). Homosexual behavior in cross-cultural perspective. *Homosexual Behavior: A Modern Perspective.* Ed. J. Marmor. New York, Basic Books. 100–122.

Carroll, J. L., K. D. Volk and J. S. Hyde (1985). Differences between males and females in motives for engaging in sexual intercourse. *Arch. Sex. Behav.* **14:** 131–139.

Cheney, D. L., R. M. Seyfarth and B. B. Smuts (1986). Social intelligence and the evolution of the primate brain. *Science* **243**: 1361–1366.

Chevalier-Skolnikoff, S. (1974). Male-female, female-female, and male-male sexual behavior in the stumptail monkey, with special attention to the female orgasm. *Arch. Sex. Behav.* **3**: 95–116.

Chevalier-Skolnikoff, S. (1975). Heterosexual copulatory patterns in stumptail macaques *(Macaca arctoides)* and in other macaque species. *Arch. Sex. Behav.* **4**: 119–220.

Chevalier-Skolnikoff, S. (1976). Homosexual behavior in a laboratory group of stumptail monkeys *(Macaca arctoides)*. *Arch. Sex. Behav.* **5**: 511–528.

Clutton-Brock, T. and P. H. Harvey (1976). Evolutionary rules and primate societies. *Growing Points in Ethology.* Ed. P. P. G. Bateson and R. A. Hinde. Cambridge, Cambridge University Press. 195–237.

Cohen, J. (1977). *Reproduction.* London, Butterworths.

Collins, J. A., W. Wrixon, L. B. Janes and E. H. Wilson (1983). Treatment—independent pregnancy among infertile couples. *New Eng. J. Med.* **309**: 1201–1206.

Control, C. f. D. (1990). Estimates of HIV prevalence and projected AIDS cases: Summary of a workshop, Oct. 31–Nov. 1, 1989. *Morb. and Mort. Week. Report* **39**: 110–119.

Control, C. f. D. (1991). Summary of notable disease, United States 1990. *Morb. and Mort. Week. Report* **39**: 16.

Crockett, C. M. (1984). Emigration of red howler monkeys and the case for female competition. *Female Primates; Studies by Woman Primatologists.* Ed. M. F. Small. New York, Alan R. Liss. 159–173.

Cunningham, M. R., A. P. Barbee and C. L. Pike (1990). What do women want? Facial metric assessment of multiple mates in the perception of male facial attractiveness. *J. Personal. and Social Psychol.* **59**: 61–72.

Cutler, W. B., C. R. Garcia and A. M. Krieger (1979). Sexual behavior frequency and menstrual cycle length in mature premenopausal women. *Psychoneuroendo.* **4**: 297–309.

Cutler, W. B., C. R. Garcia and A. M. Krieger (1980). Sporadic sexual behavior and menstrual cycle length in women. *Horm. and Behav.* **14**: 163–172.

Cutler, W. B., G. Preti, G. R. Huggins, B. Erickson and C. R. Garcia (1985). Sexual behavior frequency and biphasic ovulatory type menstrual cycles. *Physiol. Behav.* **34**: 805–816.

Cutler, W. B., G. Preti, A. M. Krieger, G. R. Huggins, C. R. Garcia and H. J. Lawley (1986). Human axillary secretions influence women's menstrual cycles: The role of donor extract from men. *Horm. and Behav.* **20**: 463–473.

Daniels, D. (1983). The evolution of concealed ovulation and self-deception. *Ethol. and Sociobiol.* **4**: 69–87.

Darling, C. A., J. K. Davidson and C. Conway-Welch (1990). Female ejaculation: Perceived origins, the Graafenberg spot/area, and sexual responsiveness. *Arch. Sex. Behav.* **19**: 29–47.

Darling, C. A., J. K. Davidson and D. A. Jennings (1991). The female sexual response revisited: Understanding the multiorgasmic experience in women. *Arch. Sex. Behav.* **20**: 527–540.

Darwin, C. (1859). *The Origin of Species by Means of Natural Selection*. London, John Murray.

Darwin, C. (1871). *The Descent of Man, and Selection in Relation to Sex*. London, John Murray.

Davis, K. B. (1929). *Factors in the Sex Life of Twenty-two Hundred Women*. New York, Harper and Brothers.

Dawkins, R. (1976). *The Selfish Gene*. Oxford, Oxford University Press.

Deaux, K. and R. Hanna (1984). Courtship in the personals column: The influence of gender and sexual orientation. *Sex Roles* **11**: 363–379.

Dewsbury, D. A. (1982). Ejaculate cost and male choice. *Amer. Nat.* **119**: 601–610.

Di Fiore, A. and D. Rendall (1994). Evolution of social organization; A reappraisal for primates by using phylogenetic methods. *Proc. Nat. Acad.* **91**: 9941–9945.

Diamond, J. (1992). *The Third Chimpanzee*. New York, HarperCollins.

Diamond, J. M. (1986). Ethnic differences: Variations in human testes size. *Nature* **320**: 488–489.

Diamond, M. (1993). Homosexuality and bisexuality in different populations. *Arch. Sex. Behav.* **22**: 291–310.

Dickemann, M. (1979). The ecology of mating systems in hypergynous dowry societies. *Soc. Sci. Informat.* **18**: 163–195.

Dienske, H. (1986). A comparative approach to the question of why human infants develop so slowly. *Primate Ontogeny, Cognition, and Social Behavior*. Ed. J. G. Else and P. C. Lee. Cambridge, University of Cambridge Press. 147–154.

Dion, K. K. (1981). Physical attractiveness, sex roles, and heterosexual attraction. *The Bases of Human Sexual Attraction*. Ed. M. Cook. New York, Academic Press.

Dixon, A. F. (1983). Observations on the evolution and behavioral significance of "sexual skin" in female primates. *Advan. Stud. Behav.* **13**: 63–106.

Döring, C. H., H. C. Kraemer, H. K. H. Brodies and D. Hamburg (1975). A cycle of plasma testosterone in the human male. *J. Clin. Endo. and Metabol.* **40**: 492–500.

Dörner, G. (1976). *Hormones and Brain Differentiation*. Amsterdam, Elsvier.

Dörner, G. (1988). Neuroendocrine response to estrogen and brain differentiation in heterosexuals, homosexuals, and transsexuals. *Arch. Sex. Behav.* **17**: 57–75.

Dörner, G., W. Rhode, F. Stahl, L. Krell and W. G. Masius (1975). A neuroendocrine predisposition for homosexuality in men. *Arch. Sex. Behav.* **4**: 1–8.

Doty, R. L. (1981). Olfactory communication in humans. *Chem. Senses* **6**: 351–376.

Duck, S. and D. Miell (1983). Mate choice in humans as an interpersonal process. *Mate Choice*. Ed. P. Bateson. Cambridge, Cambridge University Press. 377–386.

Duck, S. W. (1973). *Personal Relationships and Personal Constructs*. New York, John Wiley and Sons.

Eckland, B. K. (1968). Theories of mate selection. *Soc. Biol.* **15**: 71–97.

Ehrhardt, A. A. and H. F. L. Meyer-Bahlberg (1981). Effects of prenatal sex hormones on gender-related behavior. *Science* **211**: 1312–1318.

Eisenberg, J. F. (1981). *The Mammalian Radiation*. Chicago, University of Chicago Press.

Elke, D. E., T. J. Bouchard, J. Bohlen and L. L. Heston (1986). Homosexuality in monozygotic twins reared apart. *Brit. J. Psych.* **148**: 421–425.

Ellis, B. J. (1992). The evolution of sexual attraction: Evaluative mechanisms in women. *The Adapted Mind; Evolutionary Psychology and the Generation of Culture*. Ed. J. H. Barkow, L. Cosmides and J. Tooby. Oxford, Oxford University Press. 267–288.

Ellis, L., M. A. Ames, W. Peckham and D. Burke (1988). Sexual orientation of human offspring may be altered by severe maternal stress during pregnancy. *J. Sex. Res.* **25**: 152–157.

Elmer-Dewitt, P. (1994). Orgies on-line. *Time*, **May 31**: 61.

Ezzel, C. (1991). Brain feature linked to sexual orientation. *Science News* **140**: 143.

Fallon, A. E. and P. Rozin (1985). Sex differences in perception of desirable body shape. *J. Abnormal Psych.* **94**: 102–105.

Fay, R. E., C. F. Turner, A. D. Klassen and J. H. Gagnon (1989). Prevalence and patterns of some gender sexual contact among men. *Science* **243**: 343–348.

Fedigan, L. M. (1986). The changing role of women in models of human evolution. *Ann. Rev. Anthropol.* **15**: 25–66.

Fedigan, L. M. and H. Gouzoules (1978). The consort relationship in a troop of Japanese monkeys. *Recent Advances in Primatology*. Ed. D. J. Chivers and J. Herbert. New York, Academic Press. 493–495.

Fisher, H. (1982). *The Sex Contract*. New York, William Morrow.

Fisher, H. E. (1989). Evolution of human serial pair bonding. *Amer. J. Phys. Anthropol.* **78**: 331–354.

Fisher, H. E. (1992). *Anatomy of Love*. New York, W. W. Norton.

Fleagle, J. G. (1988). *Primate Adaptation and Evolution*. New York, Academic Press.

Foley, R. (1987). *Another Unique Species; Patterns in Human Evolutionary Ecology*. London, Longman Scientific and Technical.

Ford, C. A. and F. A. Beach (1951). *Patterns of Sexual Behavior*. New York, Harper and Brothers.

Fox, A. (1972). Studies on the relationship of plasma testosterone and human activity. *J. Endocrinol.* **52**: 51–85.

Fox, C. A. (1967). Uterine sucking during orgasm. *Brit. Med. J.* **1**: 300.

Fox, C. A., H. S. Wolff and J. A. Baker (1970). Measurement of intra-vaginal and intra-uterine pressure during human coitus by radio-telemetry. *J. Repro. Fert.* **22**: 243–251.

Fox, R. (1980). *The Red Lamp of Incest*. New York, E. P. Dutton.

France, A. d. C. S. e. (1992). AIDS and sexual behavior in France. *Nature* **360**: 407–409.

Frayser, S. G. (1985). *Varieties of Sexual Experience: An Anthropological Perspective on Human Sexuality*. New Haven, HRAF Press.

Freud, S. (1905). *Three Essays on the Theory of Sexuality*. London, Hogarth Press.

Freund, M. (1963). Effect of frequency of emission on semen output and an estimate of daily sperm production in man. *J. Reprod. Fert.* **6**: 269–286.

Freundl, G., H. J. Grimm and N. Hormann (1988). Selective filtration of abnormal spermatozoa by the cervical mucus. *Hum. Repro.* **3**: 277–280.

Gallup, G. G. (1982). Permanent breast enlargement in human females: A sociobiological analysis. *J. Human Evol.* **11**: 597–601.

Gallup, G. G. and S. D. Suarez (1983). Optimal reproductive strategies for bipedalism. *J. Human Evol.* **12**: 193–196.

Gaulin, S. J. L. and A. Schlegel (1980). Paternal confidence and paternal investment: A cross-cultural test of a sociobiological hypothesis. *Ethol. and Sociobiol.* **1**: 301–309.

Gebhard, P. H. (1966). Factors in marital orgasm. *J. Soc. Issues* **22**: 88–95.

Gebhard, P. H. and A. B. Johnson (1979). *The Kinsey Data: Marginal Tabulations of the 1938–1963 Interviews Conducted by the Institute for Sex Research.* Philadelphia, W. H. Saunders.

Ghiglieri, M. P. (1987). Sociobiology of the great apes and the hominid ancestor. *J. Human Evol.* **16**: 319–357.

Gladue, B. A. and H. J. Delaney (1990). Gender differences in perception of attractiveness in men and women in bars. *Personal. and Soc. Psychol. Bull.* **16**: 378–391.

Gladue, B. A., R. Green and R. E. Hellman (1984). Neuroendocrine response to estrogens and sexual orientation. *Science* **225**: 1496–1499.

Goodall, J. (1986). *The Chimpanzees of Gombe.* Cambridge, Harvard University Press.

Goodfoot, D. A., H. Westerborg-van Loon, W. Groenveld and A. Koos Slob (1980). Behavioral and physiological evidence of sexual climax in the female stumptail macaque *(Macaca arctoides)*. *Science* **208**: 1477–1479.

Gooren, L. (1986). The neuroendocrine response of leutinizing hormone to estrogen administration in the human is not sex specific but dependent on the hormonal environment. *J. Clin. Metab.* **63**: 589–593.

Gooren, L. (1990). Biomedical theories of sexual orientation: A critical examination. *Homosexuality/Heterosexuality: Concepts of Sexual Orientation.* Ed. D. P. McWhirter, S. A. Sanders and J. M. Reinisch. New York, Oxford University Press. 71–87.

Gorski, R. A., J. H. Gordon, J. E. Shryne and A. M. Southam (1978). Evidence for a morphological sex difference in the medial preoptic area of the rat brain. *Brain Res.* **148**: 333–346.

Gould, S. J. (1987). Freudian slip. *Nat. Hist.* **April:** 15–21.

Gouzoules, H. and R. W. Goy (1983). Physiological and social influences on mounting behavior of troop living female monkeys *(Macaca fuscata)*. *Amer. J. Primatol.* **5**: 39–49.

Graham, C. A. (1991). Menstrual synchrony: An update and review. *Human Nat.* **2**: 293–311.

Graham, C. A. and W. C. McGrew (1980). Menstrual synchrony in female undergraduates living on a coeducational campus. *Psychoneuroendo.* **5**: 245–252.

Greer, J. H., P. Morokoff and P. Greenwood (1974). Sexual arousal in women: the development of a measurement device for vaginal blood flow volume. *Arch. Sex. Behav.* **3**: 559–564.

Gregor, T. (1985). *Anxious Pleasures: The Sexual Lives of an Amazonian People.* Chicago, University of Chicago Press.

Greiling, H. (1993). Women's Short-term Sexual Strategies Conference. *Evolution and the Human Sciences,* London, London School of Economics Centre for the Philosophy of the Natural and Social Sciences.

Grmek, M. D. (1990). *History of AIDS: Emergence and Origins of a Modern Pandemic.* Princeton, Princeton University Press.

Guthrie, R. D. (1970). Evolution of human threat display organs. *Evol. Biol.* **4**: 257–302.

Hahn, H. and R. Stout (1994). *The Internet Complete Reference*. New York, Osborne–McGraw-Hill.

Hamer, D. H., S. Hu, V. L. Magnuson, N. Hu and M. L. Pattatucci (1993). A linkage between DNA markers on the X chromosomes and male sexual orientation. *Science* **261**: 321–327.

Hamilton, W. J. and P. C. Arrowood (1978). Copulatory vocalization of chacma baboons *(Papio ursinus)*, gibbons *(Hylobates hoolock)*, and humans. *Science* **200**: 1405–1409.

Harcourt, A. H. (1989). Deformed sperm are probably not adaptive. *Animal Behav.* **37**: 863–865.

Harcourt, A. H. (1991). Sperm competition and the evolution of nonfertilizing sperm in mammals. *Evol.* **45**: 314–328.

Harcourt, A. H., P. H. Harvey, S. G. Larson and R. V. Short (1981). Testis weight, body weight and breeding systems in primates. *Nature* **293**: 55–57.

Harrison, A. A. and L. Saeed (1977). Let's make a deal: An analysis of revelations and stipulations in lonely hearts advertisements. *J. Personal. and Soc. Psych.* **35**: 257–264.

Hart, B. L. and M. G. Leedy (1985). Neurological basis of male sexual behavior: A comparative analysis. *Handbook of Behavioral Neurobiology*. Ed. N. Adler, R. W. Goy and D. W. Pfaff. New York, Plenum Press. 373–422.

Hart, C. W. M., A. R. Pilling and J. C. Goodale (1988). *The Tiwi of Northern Australia*. Fort Worth, Holt, Rinehart and Winston.

Harvey, P. H. and A. H. Harcourt (1984). Sperm competition, testes size, and breeding systems in primates. *Sperm Competition and the Evolution of Animal Mating Systems*. Ed. P. L. Smith. New York, Academic Press. 589–600.

Harvey, P. H. and R. M. May (1989). Out for the sperm count. *Nature* **337**: 508–509.

Harvey, S. M. (1987). Female sexual behavior: Fluctuations during the menstrual cycle. *J. Psychosomat. Med.* **31**: 101–110.

Hein, M. (1993). *The Metaphysics of Virtual Reality*. New York, Oxford University Press.

Herdt, G. H. (1981). *Guardians of the Flutes: Idioms of Masculinity*. New York, McGraw-Hill.

Heston, L. L. and J. Shields (1968). Homosexuality in twins. *Arch. Gen. Psych.* **18**: 149–160.

Hickling, E. J., R. C. Nocel and F. D. Yutgles (1979). Attractiveness and occupational status. *The Psychol.* **102**: 71–76.

Hite, S. (1976). *The Hite Report*. New York, Dell.

Hooker, E. (1965). Male homosexuals and their "worlds." *Sexual Inversions*. Ed. J. Marmor. New York, Basic Books. 83–107.

Hoon, P. W., K. Bruce and B. Kinchole (1982). Does the menstrual cycle play a role in sexual arousal? *Psychophysiology* **19**: 21–26.

Howard, J. A., P. Blumenstein and P. Schwartz (1987). Sociological or evolutionary theory? Some observations on preferences in human mate selection. *J. Personal. and Soc. Psych.* **53**: 194–200.

Hunt, M. (1974). *Sexual Behavior in the 1970s*. Chicago, Playboy Press.

Imperato-McGinley, J., R. E. Peterson, G. Teofilo and E. Sturla (1979). Androgens and the evolution of male-gender identity among pseudohermaphrodites with 5-alph-reductase deficiency. *New Eng. J. Med.* **300**: 1233–1237.

Insler, V. and B. Luenfled (1993). *Infertility: Male and Female.* Edinburgh, Churchill Livingstone.

Irons, W. G. (1983). Human female reproductive strategies. *Social Behavior of Female Vertebrates.* Ed. S. W. Wasser. New York, Academic Press. 169–213.

Isay, R. A. (1989). *Being Homosexual; Gay Men and Their Development.* New York, Farrar Straus and Giroux.

Isbell, L. A. and T. P. Young (1995). Bipedalism and reduced group size: Alternative evolutionary responses to decreased resource availability. *J. Human Evol.:* in press.

Jackson, L. (1992). *Physical Appearance and Gender: Sociobiological and Sociocultural Perspectives.* Albany, State University of New York Press.

James, W. H. (1971). The distribution of coitus within the human intermenstruum. *J. Biosoc. Sci.* **3:** 159–171.

Jankowiak, W. R., E. M. Hill and J. M. Donovan (1992). The effect of sex and sexual orientation on attractiveness judgments: An evolutionary interpretation. *Ethol. and Sociobiol.* **13:** 73–85.

Janus, S. S. and C. L. Janus (1993). *The Janus Report on Sexual Behavior.* New York, John Wiley and Sons.

Jarett, L. R. (1984). Psychosocial and biological influences on menstruation: Synchrony, cycle length, and regularity. *Psychoneuroendo.* **9:** 21–28.

Jöchle, W. (1975). Current research in coitus-induced ovulation. *J. Repro. Fert. Supp.* **22:** 165–207.

Johnson, A. M., J. Wadsworth, K. Wellig, S. Bradshaw and J. Field (1992). Sexual lifestyles and HIV risk. *Nature* **360:** 410–412.

Johnson, V. S. and M. Franklin (1993). Is beauty in the eye of the beholder? *Ethol. and Sociobiol.* **14:** 183–199.

Jolly, A. (1986). *The Evolution of Primate Behavior.* New York, Macmillan.

Kallmann, F. J. (1952). Comparative twin study on genetic aspects of male homosexuality. *J. Nerv. Ment. Dis.* **115:** 283–298.

Kano, T. (1992). *The Last Ape; Pygmy Chimpanzee Behavior and Ecology.* Palo Alto, Stanford University Press.

Kaplan, H. (1979). *Disorders of Sexual Desire.* New York, Harper and Row.

Kenagy, G. J. and S. C. Trombulak (1986). Size and function of mammalian testes in relation to body size. *J. Mammal.* **67:** 1–22.

Kendrick, D. T., G. E. Groth, M. R. Trost and E. K. Sadalla (1993). Integrating evolutionary and social exchange perspectives on relationships: Effects of gender, self-appraisal, and involvement levels of mate selection criteria. *J. Personal. and Soc. Psychol.* **64:** 951–969.

Kendrick, D. T., E. K. Sadalla, G. Groth and M. R. Trost (1990). Evolution, traits, and the stages of human courtship: Quantifying the parental investment model. *J. Personal.* **58:** 97–116.

Kendrick, K. M. and A. F. Dixson (1986). Anteromedial hypothalamic lesions block proceptivity but not receptivity in the female common marmoset *(Callithrix jaccus). Brain Res.* **375:** 221–229.

Kiltie, R. A. (1982). On the significance of menstrual synchrony in closely associated women. *Amer. Nat.* **119:** 414–419.

Kimura, D. (1992). Sex differences in the brain. *Sci. Amer.*, **September:** 119–125.

Kinsey, A. C., W. B. Pomeroy and C. E. Martin (1948). *Sexual Behavior in the Human Male.* Philadelphia, W. B. Saunders.

Kinsey, A. C., W. B. Pomeroy, C. E. Martin and P. H. Gebhard (1953). *Sexual Behavior in the Human Female.* Philadelphia, W. B. Saunders.

Kluckholn, C. (1948). An anthropologist views it. *Sex Habits of American Men: A Symposium on the Kinsey Report.* Ed. A. Deutch. New York, Prentice Hall. 88–104.

Konner, M. (1990). *Why the Reckless Survive and Other Secrets of Human Nature.* New York, Penguin Books.

Lancaster, J. B. and C. S. Lancaster (1983). Parental investment: The hominid adaptation. *How Humans Adapt.* Ed. D. J. Ortner. Washington, D.C., Smithsonian Institution Press. 33–65.

Lancaster, J. B. and C. S. Lancaster (1987). The watershed; Change in parental-investment and family formation strategies in the course of human evolution. *Parenting Across the Lifespan; Biosocial Dimensions.* Ed. J. B. Lancaster, J. Altmann, A. S. Rossi and L. K. Sherrod. New York, Aldine de Gruyter. 187–205.

Lange, J. D., W. A. Brown, J. P. Wincze and W. Zwick (1980). Serum testosterone concentrations and penile tumescence changes in men. *Horm. and Behav.* **14:** 267–270.

Langlois, J. H. and L. A. Roggman (1990). Attractive faces are only average. *Psychol. Sci.* **1:** 115–121.

Lefrançois, G. R. (1993). *The Lifespan.* Belmont, CA, Wadsworth Publishing Company.

LeVay, S. (1991). A difference in hypothalamic structure between heterosexual and homosexual men. *Science* **253:** 1034–1037.

LeVay, S. and D. H. Hamer (1994). Evidence for a biological influence in male homosexuality. *Sci. Amer.* **270:** 44–49.

Lévi-Strauss, C. (1969). *The Elementary Structures of Kinship.* Boston, Beacon Press.

Levin, R. J. (1981). The female orgasm—a current appraisal. *J. Psychosomat. Res.* **25:** 119–133.

Lovejoy, O. C. (1981). The origin of man. *Science* **211:** 341–350.

Low, B. S., R. D. Alexander and K. M. Noonan (1987). Human hips, breasts and buttocks: Is fat deceptive? *Ethol. and Sociobiol.* **8:** 249–257.

MacFarland, L. Z. (1976). Comparative anatomy of the clitoris. *The Clitoris.* Ed. T. P. Lowry and T. S. Lowry. St. Louis, W. H. Green, Inc. 22–34.

Malinowski, B. (1965). *Sex and Repression in Savage Society.* New York, World Publishing Company.

Mann, T. and C. Lutwak-Mann (1981). *Male Reproductive Function and Semen.* Berlin, Springer-Verlag.

Margulis, L. and D. Sagan (1985). *The Origins of Sex.* New Haven, Yale University Press.

Margulis, L. and D. Sagan (1988). Sex; The cannibalistic legacy of primordial androgynes. *The Evolution of Sex.* Ed. R. Bellig and G. Stevens. San Francisco, Harper and Row. 23–48.

Marshall, D. S. and R. C. Suggs (1971). *Human Sexual Behavior; Variations in the Ethnographic Spectrum.* New York, Basic Books.

Martin, E. (1987). *The Woman in the Body.* Boston, Beacon Press.

Martin, R. D. (1990). *Primate Origins and Evolution; A Phylogenetic Reconstruction.* Princeton, Princeton University Press.

Mascia-Lees, F. F., J. H. Relethford and T. Sorger (1986). Evolutionary perspectives on permanent breast enlargement in human females. *Amer. Anthropol.* **88**: 423–428.

Mason, W. (1966). Social organization of the South American monkey *Callicebus moloch:* A preliminary report. *Tulane Stud. Zool.* **13**: 23–28.

Masters, W. and V. Johnson (1965a). The sexual response cycle of the human female: I. Gross anatomical considerations. *Sex Research: New Developments.* Ed. J. Money. New York, Holt, Rinehart and Winston. 53–89.

Masters, W. and V. Johnson (1965b). The sexual response cycle of the human female: II. The clitoris: Anatomical and clinical considerations. *Sex Research: New Developments.* Ed. J. Money. New York, Holt, Rinehart, and Winston. 90–104.

Masters, W. and V. Johnson (1966). *Human Sexual Response.* Boston, Little, Brown and Company.

Matteo, S. and E. F. Rissman (1984). Increased sexual activity during the midcycle portion of the human menstrual cycle. *Horm. and Behav.* **18**: 249–255.

Maynard Smith, J. (1971). What use is sex? *J. Theor. Biol.* **30**: 319–335.

Maynard Smith, J. (1978). *The Evolution of Sex.* Cambridge, Cambridge University Press.

Maynard Smith, J. (1988). The evolution of sex. *The Evolution of Sex.* Ed. R. Bellig and G. Stevens. San Francisco, Harper and Row. 3–20.

McCance, R. A., M. C. Luff and E. E. Widdowson (1937). Physical and emotional periodicity in women. *J. Hyg.* **37**: 571–611.

McClintock, M. (1971). Menstrual synchrony and suppression. *Nature* **229**: 244–245.

McHenry, H. M. (1991). Sexual dimorphism in *Australopithecus afarensis. J. Human Evol.* **20**: 21–32.

McHenry, H. M. (1992). Body size and proportions in early hominids. *Amer. J. Physical Anthropol.* **87**: 405–431.

Meuwissen, I. and R. Over (1992). Sexual arousal across phases of the human menstrual cycle. *Arch. of Sex. Beh.* **21**: 101–119.

Meuwissen, I. and R. Over (1993). Female sexual arousal and the Law of Initial Value: Assessment of several phases of the menstrual cycle. *Arc. Sex. Behav.* **22**: 403–413.

Meyer-Bahlberg, H. F. L. (1984). Psychoendocrine research on sexual orientation; Current status and future options. *Prog. Brain Res.* **61**: 375–398.

Michael, R. P. (1972). Determinants of primate reproductive behavior. *Acta Endocrinol. Supp.* **166**: 322–361.

Michael, R. P., R. W. Bonsall and P. Warner (1974). Human vaginal secretions: Volatile fatty acid content. *Science* **186**: 1217–1219.

Michael, R. P. and E. B. Keverne (1970). Primate sex pheromones of vaginal origin. *Nature* **223**: 84–85.

Michael, R. P., E. B. Keverne and R. W. Bonsall (1971). Pheromones: Isolation of male sex attractants from a female primate. *Science* **172**: 964–966.

Michael, R. T., J. H. Gagnon, E. O. Laumann and G. Kolata (1994). *Sex in America.* New York, Little, Brown and Company.

Miller, H. G., C. F. Turner and L. E. Moses (1990). *AIDS; The Second Decade.* Washington, D.C., National Academy Press.

Mitani, J. C. (1985). Gibbon song duets and intergroup spacing behavior. *Behaviour* **92:** 59–96.

Møller, A. P. (1988). Ejaculate quality, testes size and sperm competition in primates. *J. Human Evol.* **17:** 479–488.

Møller, A. P. (1989). Ejaculate quality, testes size and sperm production in mammals. *Funct. Ecol.* **3:** 91–96.

Money. J. (1988). *Gay, Straight, and In-Between; The Sexology of Erotic Orientation.* New York, Oxford University Press.

Money, J. (1991). The transformation of sexual terminology. *SIECUS Rep.* **19:** 10–13.

Money, J. and A. A. Ehrhardt (1972). *Man & Woman & Boy & Girl.* Baltimore, Johns Hopkins University Press.

Moore, K. L. (1988). *The Developing Human; Clinically Oriented Embryology.* Philadelphia, W. B. Saunders Co.

Morrell, M. J., J. M. Dixen, S. Carter and J. M. Davidson (1984). The influence of age and cycling status on sexual arousability in women. *Amer. J. Obst. and Gyn.* **148:** 66–71.

Morris, D. (1967). *The Naked Ape.* New York, McGraw-Hill.

Morris, N. M. and J. R. Udry (1977). A study of the relationship between coitus and the leutinizing hormone surge. *Fert. and Ster.* **28:** 440–442.

Morris, N. M. and J. R. Udry (1978). Pheromonal influences on human sexual behavior. *J. Biosoc. Sci.* **10:** 147–157.

Morris, N. M., J. R. Udry, F. Khan-Dawood and M. Y. Dawood (1987). Marital sex frequency and midcycle female testosterone. *Arch. Sex. Behav.* **16:** 27–37.

Muir, J. G. (1993). Homosexuals and the 10% fallacy. *Wall St. J.,* **March 31:** 1.

Murstein, B. I. (1976). *Who Will Marry Whom? Theories and Research in Marital Choice.* New York, Springer-Verlag.

Murstein, B. I. (1981). Process, filter, and stage theories of attraction. *The Bases of Human Sexual Attraction.* Ed. M. Cook. New York, Academic Press. 179–211.

Murstein, B. I. and P. Christy (1976). Physical attractiveness and marriage adjustment in middle aged couples. *J. Personal. and Soc. Psychol.* **34:** 537–542.

Nadler, R. (1990). Homosexuality in nonhuman primates. *Homosexuality/Heterosexuality; Concepts of Sexual Orientation.* Ed. O. P. McWhirter, S. A. Sanders and J. M. Reinisch. New York, Oxford University Press. 138–170.

Nanda, S. (1986). The hijras of India: Cultural and individual dimensions of an institutionalized third gender role. *Anthropology and Homosexual Behavior.* Ed. E. Blackwood. New York, Haworth Press. 35–54.

Nanda, S. (1991). Deviant careers: the hijras of India. *Deviance; Anthropological Perspectives.* Ed. M. Freilich, D. Raybeck and J. Savishinsky. New York, Bergin and Garvey. 149–171.

Nevid, J. S. (1984). Sex differences in factors or romantic attraction. *Sex Roles* **11:** 401–411.

Nevid, J. S. (1993). *201 Things You Should Know About AIDS and Other Sexually Transmitted Diseases.* Boston, Allyn and Bacon.

Palombit, R. (1992). Pair bonds and monogamy in wild siamang *(Hylobates syndacty-*

lus) and white-handed gibbons *(Hylobates lar)* in northern Sumatra. Davis, Calif., University of California.

Parker, G. A. (1982). Why are there so many tiny sperm? *J. Theor. Biol.* **96:** 281–294.

Parker, G. A. (1984). Sperm competition. *Sperm Competition and Animal Mating Systems.* Ed. R. L. Smith. New York, Academic Press. 1–60.

Parks, A. S. (1976). *Patterns of Sexuality and Reproduction.* Oxford, Oxford University Press.

Perrett, D. I., K. A. May and S. Yoshikawa (1994). Facial shape and judgments of female attractiveness. *Nature* **368:** 239–242.

Persky, H. (1987). *Psychoendocrinology of Human Sexual Behavior.* New York, Praeger.

Pfeffer, N. (1993). *The Stork and the Syringe: A Political History of Reproductive Medicine.* Cambridge, Blackwell Publishers.

Pillard, R. C. and J. D. Weinrich (1986). Evidence of a familial nature of male homosexuality. *Arch. Gen. Psych.* **43:** 806–812.

Pool, R. (1993). Evidence of homosexuality gene. *Science* **261:** 291–292.

Preti, G., W. B. Cutler, C. R. Garcia, G. R. Huggins and H. J. Lawley (1986). Human axillary secretions influence women's menstrual cycles: The role of donor extract of females. *Hormones and Behavior* **20:** 474–482.

Price, R. A. and S. G. Vandengerg (1979). Matching for physical attractiveness in married couples. *Personal. and Soc. Psychol. Bull.* **5:** 398–400.

Profet, M. (1993). Menstruation as a defense against pathogens transported by sperm. *Quart. Rev. Biol.* **68:** 335–386.

Pusey, A. and C. Packer (1987). Dispersal and philopatry. *Primate Societies.* Ed. B. B. Smuts, D. L. Cheney, R. M. Seyfarth, R. W. Wrangham and T. T. Strusaker. Chicago, Chicago University Press. 250–266.

Quadagno, D. M., H. E. Shubeita, J. Deck and D. Francoeur (1981). Influence of male social contacts, exercise and all-female living conditions on the menstrual cycle. *Psychoneuroendo.* **6:** 239–244.

Quiatt, D. and J. Everett (1982). How can sperm competition work? *Amer. J. Primatol. Supp.* **1:** 161–169.

Ralt, D., M. Goldenberg, P. Fetterolf, D. Thompson, J. Dor, S. Mashiach, D. Garbers and M. Eisenbach (1991). Sperm attraction to a follicular factor(s) correlates with human egg fertilizability. *Proc. Nat. Acad. Sci.* **88:** 2840–2844.

Rancourt-Laferriere, D. (1983). Four adaptive aspects of the female orgasm. *J. Biol. Soc. Struct.* **6:** 319–333.

Richard, A. F. (1987). Malagasy prosimians: Female dominance. *Primate Societies.* Ed. B. B. Smuts, D. L. Cheney, R. M. Seyfarth, R. W. Wrangham and T. T. Strusaker. Chicago, University of Chicago Press. 25–33.

Robertiello, R. C. (1970). The "clitoral versus vaginal orgasm" controversy and some of its ramifications. *J. Sex. Res.* **6:** 307–311.

Rodman, P. S. and H. M. McHenry (1980). Bioenergetics and the origin of hominid bipedalism. *Amer. J. Phys. Anthropol.* **52:** 103–106.

Rosenblatt, P. C. and R. M. Anderson (1981). Human sexuality in cross-cultural perspective. *The Basis of Human Sexual Attraction.* Ed. M. Cook. New York, Academic Press. 215–250.

Rozin, P. and A. Fallon (1988). Body image, attitudes to weight, and misperceptions of figure preference of the opposite sex: A comparison of men and women in two generations. *J. Abnormal Psychol.* **97**: 342–345.

Rubinsky, H., D. A. Eckerman and E. W. Rubinsky (1987). Early-phase physiological response patterns to psychosexual stimuli: Comparison of male and female patterns. *Arc. Sex. Behav.* **16**: 45–56.

Rushton, J. P. (1988). Genetic similarity, mate choice, and fecundity in humans. *Ethol. and Sociobiol.* **9**: 329–333.

Russell, M. J. (1976). Human olfactory communication. *Nature* **260**: 520–522.

Russell, M. J., G. M. Switz and K. Thompson (1980). Olfactory influences on the human menstrual cycle. *Pharmacol. Biochem. Behav.* **13**: 737–738.

Rutberg, A. T. (1983). The evolution of monogamy in primates. *J. Theor. Biol.* **104**: 93–112.

Sandowski, M. (1993). *With Child in Mind; Studies of Personal Encounters with Infertility.* Philadelphia, University of Pennsylvania Press.

Schmidt, G. (1975). Male-female differences in sexual arousal and behavior during and after exposure to sexually explicit stimuli. *Arc. Sex. Behav.* **4**: 353–365.

Schreiner-Engel, P., R. C. Schiavi and H. Smith (1981). Female sexual arousal: Relation between cognitive and genital assessments. *J. Sex and Marital Ther.* **2**: 256–267.

Shepard, G. (1987). Rank, gender, and homosexuality: Mombasa as a key to understanding sexual options. *The Cultural Construction of Sexuality.* Ed. P. Caplan. London, Tavistock Publishing. 240–270.

Sherfey, M. J. (1966). *The Nature and Evolution of Female Sexuality.* New York, Random House.

Short, R. V. (1976). The evolution of human reproduction. *Proc. Royal Soc. Lond.* **195**: 3–24.

Short, R. V. (1979). Sexual selection and its component parts, somatic and genetical selection, as illustrated by man and the great apes. *Adv. Stud. Behav.* **9**: 131–158.

Short, R. V. (1980). The origins of human sexuality. *Human Sexuality.* Ed. C. R. Austin and R. V. Short. Cambridge, Cambridge University Press. 1–33.

Short, R. V. (1981). Sexual selection in man and the great apes. *Reproductive Biology of the Great Apes.* Ed. C. E. Graham. New York, Academic Press. 319–341.

Short, R. V. (1984a). *One Medicine.* Ed. O. A. Ryder and M. L. Byrd. Berlin, Springer-Verlag.

Short, R. V. (1984b). Breast feeding. *Sci. Amer.* **250**: 35–41.

Short, R. V. (1984c). The role of hormones in sex cycles. *Hormones and Reproduction.* Ed. C. R. Austin and R. V. Short. Cambridge, Cambridge University Press. 42–72.

Sillén-Tullberg, B. and A. P. Møller (1993). The relationship between concealed ovulation and mating systems in anthropoid primates: A phylogenetic analysis. *Amer. Nat.* **141**: 1–25.

Singer, B. (1985). A comparison of evolutionary and environmental theories of erotic response: Part I. Structural features. *J. Sex. Res.* **21**: 229–257.

Singer, J. and I. Singer (1972). Types of female orgasm. *J. Sex. Res.* **8**: 255–267.

Singh, D. (1993). Adaptive significance of female physical attractiveness: Role of waist-to-hip ratio. *J. Personal. and Soc. Psychol.* **65**: 293–307.

Slater, P. J. B. (1978). *Sex Hormones and Behavior*. London, Edward Arnold.

Slob, A. K., M. Ernste and J. J. von der Werff ten Bosch (1991). Menstrual cycle phase and sexual arousability in women. *Arch. Sex. Behav.* **20**: 567–577.

Small, M. F. (1988). Female primate sexual behavior and conception; Are there really sperm to spare? *Curr. Anthropol.* **29**: 81–100.

Small, M. F. (1992a). The evolution of female sexuality and mate selection in humans. *J. Human Nat.*: in press.

Small, M. F. (1992b). Female choice in mating. *Amer. Sci.* **80**: 142–151.

Small, M. F. (1993). *Female Choices; Sexual Behavior of Female Primates*. Ithaca, Cornell University Press.

Smith, R. L. (1984). Human sperm competition. *Sperm Competition and the Evolution of Animal Mating Systems*. Ed. R. L. Smith. New York, Academic Press. 601–659.

Smith, T. W. (1991). Adult sexual behavior in 1989: Number of partners, frequency of intercourse and risk of AIDS. *Fam. Plan. Perspect.* **23**: 102–107.

Smuts, B. B. and J. M. Wantanabe (1990). Social relationships and ritualized greeting in male baboons. *Int. J. Primatol.* **11**: 147–172.

Sommer, V. (1988). Female-female mounting in langurs. *Int. J. Primatol.* **8**: 478.

Srivastava, A., C. Borries and V. Sommer (1991). Homosexual mounting in free-ranging Hanuman langurs. *Arch. Sex. Behav.* **20**: 487–516.

Stall, P., T. J. Coates, J. S. Mandel, E. S. Morales and J. L. Sorensen (1989). Behavioral factors and intervention. *The Epidemiology of AIDS: Expression, Occurrence, and Control of Human Immunodeficiency Virus Type I Infection*. Ed. R. A. Kaslow and D. P. Francis. New York, Oxford University Press. 266–281.

Stanislaw, H. and F. J. Rice (1987). Acceleration of the menstrual cycle by intercourse. *Psychophysiol.* **24**: 714–718.

Steinem, G. (1994). *Moving Beyond Words*. New York, Simon and Schuster.

Steklis, H. D. and C. H. Whiteman (1989). Loss of estrus in human evolution: Too many answers, too few questions. *Ethol. and Sociobiol.* **10**: 417–434.

Sterns, E. L., J. S. D. Winter and C. Faiman (1973). Effects of coitus on gonadotropins, prolactin, and sex steroid levels in man. *J. Clin. Endo. Metab.* **37**: 687–691.

Stoller, R. J. and G. H. Herdt (1985). Theories of origins of homosexuality. *Arch. Gen. Psych.* **42**: 399–404.

Stopes, M. C. (1918). *Married Love: A New Contribution to the Solution of Sex Difficulties*. London, G. P. Putnam and Sons.

Strassmann, B. I. (1981). Sexual selection, paternal care and concealed ovulation in humans. *Ethol. and Sociobiol.* **2**: 31–40.

Strathern, M. (1992). *Reproducing the Future*. New York, Routledge.

Strier, K. B. (1994). Myth of the typical primate. *Yrbk. Physical Anthropol.* **37**: 233–271.

Swaab, D. F. and E. Fliers (1985). A sexually dimorphic nucleus in the human brain. *Science* **228**: 1112–1114.

Swaab, D. F. and M. A. Hofman (1990). An enlarged suprachiasmatic nucleus in homosexual men. *Brain Res.* **537**: 141–148.

Symons, D. (1979). *The Evolution of Human Sexuality*. Oxford, Oxford University Press.

Symons, D. (1993). Beauty is in the Adaptations of the Beholder Conference. *Evolution and the Human Sciences*. London, London School of Economics Centre for the Philosophy of the Natural and Social Sciences.

Symons, D. and B. Ellis (1989). Human male-female differences in sexual desire. *The Sociobiology of Sexual and Reproductive Strategies*. Ed. A. E. Rasa, C. Vogel and E. Voland. London, Chapman and Hall. 131–146.

Szalay, F. S. and R. K. Costello (1991). Evolution of permanent estrus displays in hominids. *J. Human Evol.* **20**: 439–464.

Talmon, Y. (1964). Mate selection in collective settlements. *Amer. Sociol. Rev.* **29**: 491–508.

Tavris, C. and S. Sadd (1975). *The Redbook Report on Female Sexuality*. New York, Delacorte Press.

Thiessen, D., R. K. Young and R. Burroughs (1993). Lonely hearts advertisements reflect sexually demographic mating strategies. *Ethol. and Sociobiol.* **14**: 209–229.

Thompson-Handler, N., R. Malenky and N. Badrian (1984). Sexual behavior of *Pan paniscus* under natural conditions in the Lomako Forest, Equateur, Zaire. *The Pygmy Chimpanzee; Evolution, Biology, and Behavior*. Ed. R. L. Susman. New York, Plenum Press. 347–368.

Tierney, J. (1994). Porn, the low-slung engine of progress. *The New York Times*, **Section 2:** 1, 18.

Townsend, J. M. (1987). Sex differences among medical students: Effects of increasing socioeconomic status. *Arch. Sex. Behav.* **16**: 425–444.

Townsend, J. M. (1989). Mate selection criteria: A pilot study. *Ethol. and Sociobiol.* **10**: 241–253.

Townsend, J. M. and G. L. Levy (1990). Effects of potential partners' physical attractiveness and socioeconomic status on sexuality and partner selection. *Arc. Sex. Behav.* **19**: 149–164.

Treloar, A. E., R. E. Boynton, B. G. Behn and B. W. Brown (1967). Variation in the human menstrual cycle through reproductive life. *Int. J. Fert.* **12**: 77–126.

Trivers, R. L. (1972). Parental investment and sexual selection. *Sexual Selection and the Descent of Man*. Ed. B. Campbell. Chicago, Aldine. 1136–1179.

Turke, P. W. (1984). Effects of ovulatory concealment and synchrony of protohominid mating systems and parental roles. *Ethol. and Sociobiol.* **5**: 33–44.

Turner, R. (1993). Landmark French and British studies examine sexual behavior, including multiple partners, homosexuality. *Fam. Plan. Persp.* **25**: 91–92.

Turner, W. J. (1994). Comments on discordant monozygotic twinning in homosexuality. *Arch. Sex. Behav.* **23**: 115–119.

Tutin, C. E. (1979). Mating patterns and reproductive strategies in a community of wild chimpanzees *(Pan troglodytes schweinfurthii)*. *Behav. Ecol. Sociobiol.* **6**: 29–38.

Udall, J. S. and S. H. Geist (1943). Effect of androgen upon libido of women. *J. Clin. Endocrinol.* **3**: 235–238.

Udry, J. R. and N. M. Morris (1968). Distribution of coitus in the menstrual cycle. *Nature* **220**: 593–596.

Udry, J. R. and N. M. Morris (1977). The distribution of events in the human menstrual cycle. *J. Reprod. Fert.* **51**: 419–425.

Van Den Bergh, P. L. (1979). *Human Family Systems; An Evolutionary View*. New York, Elsevier.

Veith, J. L., M. Buck, S. Getzlaf, P. Van Dalfsen and S. Slade (1983). Exposure to men influences the occurrence of ovulation in women. *Physiol. and Behav.* **31:** 313–315.

Voeller, B. (1991). AIDS and heterosexual anal intercourse. *Arch. Sex. Behav.* **20:** 233–276.

Weeling, K., J. Field, J. Wadsworth, A. M. Johnson, R. M. Anderson and S. A. Bradshaw (1990). Sexual lifestyles under scrutiny. *Nature* **348:** 276–278.

Wexelblat, A. (1993). *Virtual Reality: Applications and Explorations.* Boston, Academic Press Professional.

Whitam, F. L., M. Diamond and J. Martin (1993). Homosexual orientation in twins: A report on 61 pairs and 3 triplet sets. *Arch. Sex. Behav.* **20:** 187–206.

White, T. D., G. Suwa and B. Asfaw. *Australopithecus ramidus,* a new species of early hominid from Aramis, Ethiopia. *Nature* **371:** 306–312.

White, I. G. and A. B. Kar (1973). Aspects of the physiology of sperm in the female genital tract. *Contraception* **3:** 183–194.

Whitehead, H. (1981). The bow and the burden strap: A new look at institutionalized homosexuality in Native North America. *Sexual Meanings; The Cultural Construction of Gender and Sexuality.* Ed. S. Ortner and H. Whitehead. Cambridge, Cambridge University Press. 80–115.

Whitters, W. L. and P. Jones-Whitters (1980). *Human Sexuality; A Biological Perspective.* New York, Van Nostrand.

Whyte, M. K. (1978). Cross-cultural codes dealing with the relative status of women. *Ethnology* **17:** 211–237.

Whyte, M. K. (1990). *Dating, mating, and marriage.* New York, Aldine de Gruyter.

Wiederman, M. W. and E. R. Allgeier (1992). Gender differences in mate selection criteria: Sociobiological or socioeconomic explanation. *Ethol. and Sociobiol.* **13:** 115–124.

Williams, G. C. (1975). *Sex and Evolution.* Princeton, Princeton University Press.

Williams, W. L. (1986). *The Spirit and the Flesh: Sexual Diversity in American Indian Culture.* Boston, Beacon Press.

Wilson, E. O. (1978). *On Human Nature.* Cambridge, Harvard University Press.

Wilson, G. D. (1981). Cross-generational stability of gender differences in sexuality. *Personal. and Indiv. Diff.* **2:** 254–257.

Wolfe, L. (1980). The sexual profile of the *Cosmopolitan* girl. *Cosmopolitan,* **September:** 254–265.

Wolfe, L. M. (1979). Behavioral patterns of estrous females of the Arashiyama West troop of Japanese macaques *(Macaca fuscata). Primates* **20:** 525–534.

Wolfe, L. M. (1984). Female rank and reproductive success among Arashiyama B Japanese macaques *(Macaca fuscata). Int. J. Primatol.* **5:** 133–143.

Wolfe, L. M. (1986). Sexual strategies of female Japanese macaques *(Macaca fuscata). Hum. Evol.* **1:** 267–275.

Wrangham, R. W. (1980). An ecological model of female-bonded primate groups. *Behaviour* **75:** 262–300.

Wright, R. (1994). *The Moral Animal.* New York, Pantheon Books.

Young, W. C., R. W. Goy and C. H. Phoenix (1964). Hormones and sexual behavior. *Science* **143:** 212–218.

Zaviacic, M. and B. Whipple (1993). Update on the female prostate and the phenomenon of female ejaculation. *J. Sex. Res.* **30**: 148–151.

Zimmerman, S. J., M. B. Maude and M. Moldawar (1965). Frequent ejaculation and total sperm count, motility and forms in humans. *Fert. and Ster.* **16**: 342–345.

Zumpe, D. and R. P. Michael (1968). The clutching reaction and orgasm in the female rhesus monkey *(Macaca mulatta). J. Endocrin.* **40**: 117–123.

Acknowledgments

Over the years, a number of colleagues, teachers, students, and friends have had a significant impact on how I view the path of human evolution. I thank especially Sarah Blaffer Hrdy, David Buss, Adam Clark Arcadi, Anthony Di Fiore, Lynne Isbell, Fred Lorey, Henry M. McHenry, John Mitani, Ryne Palombit, Peter S. Rodman, Becky Rolfs, and David Glenn Smith.

A network of editors, creative writers, and science writers have been more than generous during the past few years as I slid into popular writing. These individuals not only taught me the trade of writing as a career, they also welcomed me into the field with open hearts (and the names of agents, magazine editors, and contacts). Their generosity and encouragement have been the best thing that has happened to me in recent years. I thank especially Diane Ackerman, Paul Cody, Nathaniel Comfort, Ann Gibbons, Bruce Lewenstein, Roger Lewin, Jeanne Mackin, Steve Madden, Steve Minsky, Michael May, Pat Shipman, Charles H. Small, and Mark Wheeler. Maxine thanks Kenny Berkowitz, Brad Edmonson, and James J. Gould in particular.

Other friends and family deserve my thanks for being so forthcoming about their sexuality. I received some weird responses from people when I said I was writing a book about sex, but my good friends were kind enough to share their thoughts, and their personal experiences. I hope they don't mind being named here: Dilmeran Dunham, Nick Fowler, Dede Hatch, Jackie Hatton, Ann Jereb, John Norvell, Andrea Perkins, John Reis, Krysia Small, Carol Terrizzi, Tom Terrizzi, and Ute van den Bergh. Susan Childs Merrick and Frank B. Merrick graciously let me use their home as a way station on my many trips to New York City.

The production of this book was made fun, rather than a task, by several people. My agent, Anne Sibbald at Janklow and Nesbit, was behind the project from the beginning, and her enthusiasm is always encouraging. I have also had the luck to work with a fine editor at Anchor Books, Roger Scholl, and his detailed editing and important comments have made this a better book. More important, Roger's excitement about this book, his positive comments about my writing, and the fact that he gets my jokes have made the past many months of working on this book a real pleasure. I chose Dede Hatch to contribute photographs for this book because I knew she would add certain artistic, humorous, and poignant flair to the variety of subjects I cover. She surpassed even my very high expectations.

My greatest thanks goes to my significant other, Tim Merrick. On the day I started the chapter on male sexuality, Tim came home with the usual happy grin on his face. I began to pontificate about what I had read that day, illustrating my knowledge by poking various parts of his body. After a few stabs and grabs, he held on to my wrists in mid-reach and, still grinning, said, "I guess this is going to be a really difficult chapter for me, isn't it?" And it was. But he bore it with great cheer, always interested in what I had learned and what I was writing. He also allowed me to relate a rather private story about him at the beginning of Chapter Four, even though he knew complete strangers, all our friends, and even his parents, would read it.

M.F.S.
Ithaca, New York
1994

Notes

INTRODUCTION

1. Kinsey et al. 1948; Kinsey et al. 1953.
2. Masters and Johnson 1966.
3. Hite 1976.
4. Janus and Janus 1993.
5. Michael et al. 1994.
6. Morris 1967.

CHAPTER ONE

1. Foley 1987.
2. Fleagle 1988.
3. Fleagle 1988.
4. A new species, *Australopithecus ramidus*, dated at 4.4 million years ago, was recently found and named (White 1994). More complete discoveries of this new species should soon lend some clues to this period.
5. Isbell and Young 1995; Lovejoy 1981.
6. Foley 1987.
7. Foley 1987; Isbell and Young 1995; Rodman and McHenry 1980.
8. Foley 1987.
9. The hominid fossil record changes with each new discovery—as it should. For now, a good description of our fossil history can be found in any introductory textbook in biological anthropology.
10. Cheney et al. 1986.
11. Foley 1987.
12. Wrangham 1980.
13. Mitani 1985.
14. Isbell and Young 1995.
15. Rutberg 1983.
16. Ecologists place these categories on a continuous scale. Those species which produce infants that require little investment and often make a multitude of infants at once, such as insects, are called "r-selected species." At the other end of the scale are "K-selected species," such as humans, which have fewer total infants but invest more in each infant.
17. Eisenberg 1981.

18. Dienske 1986.
19. Martin 1990.
20. Alexander 1990; Alexander and Noonan 1979; Lovejoy 1981.
21. There are several versions of this scenario but they all add up about the same way—females using sex to gain male care for infants (Alexander 1990; Alexander and Noonan 1979; Fisher 1982; Lovejoy 1981; Strassmann 1981; Symons 1979; Turke 1984).
22. Lovejoy 1981.
23. Alexander and Noonan 1979; Lovejoy 1981.
24. Beach 1976.
25. Martin 1990.
26. Small 1993.
27. Small 1993.
28. Blaffer Hrdy and Whitten 1987; Small 1993.
29. Blaffer Hrdy 1983.
30. Mason 1966.
31. Palombit 1992.
32. Darwin 1859; Darwin 1871.
33. Clutton-Brock and Harvey 1976.
34. Blaffer Hrdy and Bennett 1981c.
35. McHenry 1991; McHenry 1992.
36. Goodall 1986; Tutin 1979.
37. Van Den Bergh 1979.
38. Fisher 1989.
39. Fisher 1992.
40. Janus and Janus 1993.
41. Broude and Greene 1976; Gebhard and Johnson 1979; Small 1992a; Whyte 1978; Wolfe 1980. The recent survey *Sex in America* disputes these numbers (Michael et al. 1994).
42. Broude and Green 1976.
43. Blaffer Hrdy 1981a.
44. Baker and Bellis 1989.
45. Blaffer Hrdy and Whitten 1987; Dixon 1983.
46. Blaffer Hrdy and Whitten 1987.
47. Alexander 1990; Alexander and Noonan 1979; Benshoff and Thornhill 1979; Daniels 1983; Fisher 1982; Lovejoy 1981; Strassmann 1981.
48. Alexander and Noonan 1979.
49. Blaffer Hrdy 1979; Blaffer Hrdy 1983; Sillén-Tullberg and Møller 1993.
50. Symons 1979.
51. Benshoff and Thornhill 1979; Symons 1979.
52. Burley 1979.
53. Blaffer Hrdy 1983. Szalay and Costel have created a pictorial history of the loss of female swellings (Szalay and Costello 1991). They imagine a chimpanzee-like creature with a swelling who then lost this swelling when she stood up and used bipedalism as a major mode of locomotion. As a consequence, they feel, human females gained big bottoms and pendulous breasts to signal their sexuality. Their scenario is unsubstantiated because males also have large bottoms—a result of increased need for the gluteus maximus to stabilize the body during bipedalism. In addition, breasts as a sexual signal is a cultural norm and not necessarily a biological motivator.
54. Dixon 1983.
55. Recently, some anthropologists have been trying to change the perspective that New World monkeys and apes are odd in their lack of swellings and other patterns of behavior. Instead, these researchers contend that the New World monkeys represent the basal primate stock, and the Old World monkeys, or cercopithecoids, are the specialized creatures (Di Fiore and Rendall 1994; Strier 1994).
56. Dixon 1983; Sillén-Tullberg and Møller 1993.
57. Dixon 1983.
58. Burt 1992.
59. Clutton-Brock and Harvey 1976.
60. Andelman 1987; Blaffer Hrdy and Whitten 1987; Small 1993.
61. Burt 1992.
62. Irons 1983; Lancaster and Lancaster 1987.
63. Steklis and Whiteman 1989.
64. Anderson 1988; Cant 1981; Caro 1987; Caro and Sellen 1990; Gallup 1982; Low et al. 1987; Mascia-Lees et al. 1986.
65. Caro 1987; Caro and Sellen 1990.
66. Anderson 1988.
67. Short 1984.
68. Mascia-Lees et al. 1986.
69. Caro and Sellen 1990.
70. Anderson 1988.

71. Short 1981.
72. Diamond 1992; Guthrie 1970.
73. Short 1981.
74. Alexander 1990; Ghiglieri 1987.

75. Foley 1987; Isbell and Young 1995.
76. Short 1976.
77. Short 1976.

CHAPTER TWO

1. In Chapter Six, I evaluate the current research on the possibility of a biological basis for homosexuality, including recent research on the genetics of homosexuality. Although researchers have specifically addressed the issue of homosexuality, no one has looked at heterosexuality.
2. Dawkins 1976; Wilson 1975.
3. Williams 1975.
4. Margulis and Sagan 1985; Margulis and Sagan 1988; Maynard Smith 1978; Williams 1975.
5. Maynard Smith 1971; Maynard Smith 1978; Maynard Smith 1988.
6. Bernstein, Hopf, and Michod 1988; Hamilton 1988; Hamilton et al. 1990.
7. Dawkins 1976.
8. Bancroft 1980; Whitters and Jones-Whitters 1980.
9. Bancroft 1980.
10. Whitters and Jones-Whitters 1980.
11. Kinsey et al. 1953.
12. Ford and Beach 1951.
13. Bancroft 1983.
14. Jolly 1986; Martin 1990.
15. Jolly 1986; Martin 1990.
16. Martin 1990.
17. Bancroft 1983; Martin 1990.
18. Doty 1981.
19. Russell 1976.
20. Avers 1974; Bermant and Davidson 1974.
21. Bancroft 1983.
22. Bancroft 1983.
23. Bancroft 1983.

24. Bancroft 1983.
25. Hart and Leedy 1985.
26. Kendrick and Dixson 1986.
27. Whitters and Jones-Whitters 1980.
28. There are several good texts on the neurophysiology of human sexual behavior, but in my reading I was also struck by how much they *don't* know. I believe this lack of knowledge is a result of our primitive understanding of how the brain actually works in the first place (Avers 1974; Bermant 1974; Whitters 1980).
29. Bancroft 1983.
30. Bancroft 1983.
31. Bancroft 1980.
32. Schmidt 1975.
33. Rubinsky et al. 1987.
34. Greer et al. 1974.
35. Bancroft 1983.
36. Masters and Johnson 1966.
37. Persky 1987.
38. Moore 1988.
39. Money and Ehrhardt 1972.
40. Persky 1987.
41. Bancroft 1983.
42. Whitters and Jones-Whitters 1980.
43. Bancroft 1980.
44. Michael and Keverne 1970; Michael et al. 1971.
45. Bancroft 1980; Slater 1978.
46. Fox 1972.
47. Avers 1974; Bermant and Davidson 1974.
48. Udall and Geist 1943.

CHAPTER THREE

1. Barret-Ducrocq 1991.
2. Janus and Janus 1993.
3. Sherfey 1966.
4. Trivers 1972.
5. For a discussion of female nonhuman primates and their sexual behavior see my previous book, *Female Choices; Sexual Behavior of Female Primates* (Small 1993). In that book, I write about the evolutionary theory of female sexuality and how

primate females don't really fit the model very well.

6. The information on ovulation can be found in any good text on female reproductive biology. I found Whitters (Whitters and Jones-Whitters 1980) particularly clear.
7. Short 1984.
8. Treloar et al. 1967.
9. Frayser 1985.
10. For an excellent discussion of feminist biology read Emily Martin's consciousness-raising volume on how we articulate the reproductive process (Martin 1987). Clearly, our cultural attitude about women in general influences the way we see the female role in the reproductive process. For another good example, read Margie Profet's recent article suggesting that menstruation is not a matter of ridding the body of disgusting, useless material but a positive action selected over time to rid the body of pathogens (Profet 1993).
11. Asso 1983.
12. Profet 1993.
13. Short 1976.
14. I thank Dr. John Capitanio for being such a good sport and allowing me to print this story.
15. McClintock 1971.
16. Graham 1991; Graham and McGrew 1980; Quadagno et al. 1981.
17. Graham 1991.
18. Russell et al. 1980. This study was repeated using sweat from four different donors and the same synchronization occurred (Preti et al. 1986).
19. Aliphatic acids found in the human vagina include acetic, propanoic, methylpropanoic, butanoic, methylbutanoic and methylpentanoic acids (Michael et al. 1974; Morris and Udry 1978). It seems that these acids, when distilled down and then wiped on the rear ends of infertile female rhesus monkeys, are extremely attractive to male monkeys—they think the females are in estrus (Michael 1972; Parks 1976).
20. Jarett 1984.
21. Burley 1979.
22. Kiltie 1982.
23. Veith et al. 1983.
24. Cutler et al. 1986.
25. Cutler et al. 1979; Cutler et al. 1980.
26. Cutler et al. 1985.
27. Stanislaw and Rice 1987.
28. Jöchle 1975; Morris and Udry 1977.
29. Benedek and Rubenstein 1939a; Benedek and Rubenstein 1939b.
30. Persky 1987; Sterns et al. 1973.
31. Persky 1987.
32. Bancroft et al. 1983.
33. Stopes 1918.
34. Asso 1983; Benedek 1952.
35. McCance et al. 1937.
36. Harvey 1987; Morris and Udry 1978; Udry and Morris 1968; Udry and Morris 1977.
37. James 1971.
38. Asso 1983; Bancroft et al. 1983; Morris et al. 1987. Other reproductive, not sexual, hormones such as estradiol don't show a midcycle association with heterosexual sex (Abplanalp et al. 1979). It may be that estrogen first stimulates the production of androgen, which in turn affects sexual desire.
39. Matteo and Rissman 1984.
40. Adams et al. 1978; Asso 1983; Harvey 1987; Stanislaw and Rice 1987.
41. Bancroft 1987; Persky 1987; Steklis and Whiteman 1989.
42. Davis 1929.
43. Greer et al. 1974.
44. There are at least forty studies on female arousal and the cycle, and none show conclusively that there is any pattern relative to ovulation or any other cycle stage (Hoon et al. 1982; Meuwissen and Over 1992; Meuwissen and Over 1993; Morrell et al. 1984; Schreiner-Engel et al. 1981; Slob et al. 1991).
45. Levin 1981.
46. Hamilton and Arrowood 1978.
47. Freud 1905; Sherfey 1966.
48. Sherfey 1966.
49. I've looked at several textbooks on female sexuality (Avers 1974; Bancroft 1983; Sherfey 1966; Whitters and Jones-Whitters 1980) and all of them, rely on Masters and Johnson's seminal work on human sexual responses (Masters and Johnson 1966). This book is still

the best reference for human sexual mechanics and unless otherwise noted, that is my source of the information on female orgasm. Masters and Johnson based their description of female and male sexual response on over ten thousand sexual bouts with 382 female and 312 male subjects during intercourse and masturbation. Data were compiled with the aid of intrauterine measuring devices, including a plastic penis with a camera, and more standard equipment such as electrocardiograms.

50. Masters and Johnson 1965b.
51. Some have objected to the Masters and Johnson scheme. They object to the definition of the plateau stage or to the idea that all female orgasms are basically alike in their response pattern (Kaplan 1979; Levin 1981).
52. Whitters has a lucid explanation of the female and male sexual response pattern (Whitters and Jones-Whitters 1980).
53. Bohlen et al. 1982a.
54. Darling et al. 1990; Zaviacic and Whipple 1993.
55. Whitters and Jones-Whitters 1980.
56. Bancroft 1983.
57. Masters and Johnson 1965a.
58. Butler and Peeler 1979; Robertiello 1970; Singer and Singer 1972.
59. Bohlen et al. 1982b.
60. Darling et al. 1991.
61. Levin 1981.
62. For example, artificial insemination, in which no woman experiences orgasm, let alone sexual arousal, results in conception. Also, rape, a terrifying experience far distant from sexuality, can also result in conception (Parks 1976).
63. Moore 1988.
64. Gould 1987; Symons 1979.
65. Anthropologist Linda Fedigan calls this the "coattails" theory of evolutionary ac-

quirement—one sex gets something just because the other has it, and female items seem to be most often explained in this manner (Fedigan 1986).
66. Alcock 1987.
67. Alcock 1987; Blaffer Hrdy 1979; Sherfey 1966.
68. Burton 1971; Chevalier-Skolnikoff 1974; Chevalier-Skolnikoff 1975; Chevalier-Skolnikoff 1976; Goodfoot et al. 1980; Zumpe and Michael 1968.
69. Female orgasm in nonhuman primates is covered in detail in *Female Choices*, pp. 137–143 (Small 1993).
70. MacFarland 1976.
71. Blaffer Hrdy 1979; Konner 1990; Small 1993.
72. Blaffer Hrdy 1979; Levin 1981.
73. Alcock 1980; Alcock 1987.
74. Gebhard 1966; Rancourt-Laferriere 1983.
75. Baker and Bellis 1989; Janus and Janus 1993; Kinsey et al. 1953; Wolfe 1980.
76. Masters and Johnson 1965a; Masters and Johnson 1965b.
77. Gebhard 1966; Rancourt-Laferriere 1983.
78. Gregor 1985; Levin 1981; Marshall and Suggs 1971.
79. Morris 1967.
80. Levin 1981.
81. Fox 1967; Fox et al. 1970.
82. Others disagree with this study. For example, Masters and Johnson placed radio-opaque liquid that was similar in consistency to semen close to the cervix in six women. Although these women masturbated to orgasm, there was no evidence of this liquid in their uteruses (Bancroft 1983).
83. Baker and Bellis 1993.
84. Hart et al. 1988.
85. Ralt et al. 1991.
86. Small 1988.

CHAPTER FOUR

1. Bem 1993.
2. Mann and Lutwak-Mann 1981; Moore 1988.
3. Bancroft 1983.
4. Bancroft 1983.
5. Lange, Brown et al. 1980.
6. Bancroft 1983.
7. Döring et al. 1975; Persky 1987.

8. Whitters and Jones-Whitters 1980.
9. Bancroft 1983.
10. Martin 1990.
11. Bedford 1977; Martin 1990.
12. Mann and Lutwak-Mann 1981.
13. Short 1981.
14. Bancroft 1983.
15. Masters and Johnson 1966; Bancroft 1983.
16. Bancroft 1983.
17. Masters and Johnson 1966.
18. Masters and Johnson 1966.
19. Whitters and Jones-Whitters 1980; Bancroft 1983.
20. Masters and Johnson 1966.
21. About 15 to 30 percent of seminal fluid comes from the prostate and the rest is from the other glands such as the seminal vesicle (Mann and Lutwak-Mann 1981).
22. Bermant and Davidson 1974.
23. Bohlen et al. 1980.
24. Mann and Lutwak-Mann 1981.
25. Mann and Lutwak-Mann 1981.
26. Mann and Lutwak-Mann 1981.
27. Short 1980.
28. Zimmerman et al. 1965.
29. Mann and Lutwak-Mann 1981.
30. Mann and Lutwak-Mann 1981.
31. White and Kar 1973.
32. Quiatt and Everett 1982.
33. Austin 1975.
34. Parker 1982.
35. Cohen 1977.
36. Parker 1984.
37. Dewsbury 1982.
38. Harcourt et al. 1981; Harvey and Harcourt 1984; Kenagy and Trombulak 1986; Møller 1988.
39. Mann and Lutwak-Mann 1981.
40. Smith 1984.
41. Short 1981. In another study of dead males, the estimated daily production was 207 million sperm per day (Amann and Howards 1980).
42. Freund 1963; Harvey and May 1989.
43. Amann and Howards 1980.
44. Harvey and May 1989.
45. Short 1981.
46. Freund 1963.
47. Harcourt 1991; Short 1980.
48. Short 1979; Harcourt et al. 1981; Harvey and Harcourt 1984; Møller 1988; Møller 1989.
49. Short 1976.
50. Kinsey et al. 1953; Hunt 1974; Tavris and Sadd 1975; Gebhard and Johnson 1979; Wolfe 1980; Small 1992a; Janus and Janus 1993. A high number has recently been disputed, although there are major problems in the way these data were collected (Michael et al. 1994).
51. Broude and Greene 1976; Whyte 1978; Small 1992a.
52. Broude and Greene 1976; Whyte 1978; Small 1992a.
53. Rosenblatt and Anderson 1981.
54. As Robert Smith writes, "The biological irony of the double standard is that males could not have been selected for promiscuity if historically females had always denied them opportunity for expression of the trait" (Smith 1984), p. 602.
55. Baker and Bellis 1989.
56. Diamond 1992.
57. Cited in (Smith 1984).
58. Gaulin and Schlegel 1980.
59. Mann and Lutwak-Mann 1981.
60. Mann and Lutwak-Mann 1981; Gallup and Suarez 1983; Baker and Bellis 1993a.
61. Cohen 1977.
62. Baker and Bellis 1989.
63. Baker and Bellis 1988.
64. Harcourt 1989; Harcourt 1991.
65. Freundl et al. 1988.
66. Even with body size accounted for, African blacks have larger testes and higher sperm counts than European Caucasians. Japanese and Korean males have even smaller testes and lower sperm counts. (Short 1984; Diamond 1986; Harvey and May 1989).
67. Baker and Bellis 1993b.
68. Baker and Bellis 1989; Baker and Bellis 1993b.
69. Baker and Bellis 1993b.
70. To cite just a few studies: Kinsey et al. 1953; Tavris and Sadd 1975; Wolfe 1980; Janus and Janus 1993.
71. Buss 1994; Symons 1979.
72. Symons and Ellis 1989.
73. Carroll et al. 1985 and see Symons (Symons 1979) for further elaboration.

74. Blaffer Hrdy 1981a; Small 1993.
75. Wolfe 1980; Janus and Janus 1993.
76. Hite 1976.
77. Carroll et al. 1985. This study also shows how sad some sex lives are. When asked what they do when they can't have sex, 80 percent of the men said they engage in a vigorous activity, while 50 percent of the women said they watch TV.
78. Small 1993.

CHAPTER FIVE

1. Symons 1979; Buss 1994. For an explanation of various hypotheses put forth by evolutionary psychologists, read Robert Wright's book *The Moral Animal,* written for the nonacademic audience (Wright 1994).
2. Ellis 1992.
3. Trivers 1972.
4. Buss 1994.
5. Symons 1979; Buss 1994.
6. Small 1993.
7. Eckland 1968; Duck and Miell 1983; Buss 1994.
8. Eckland 1968; Duck 1973; Duck and Miell 1983; Kenrick et al. 1993.
9. Eckland 1968.
10. Duck 1973; Murstein 1981.
11. Jackson 1992; Buss 1994.
12. Richard 1987.
13. Wrangham 1980.
14. Goodall 1986.
15. Kano 1992.
16. Small 1993.
17. Evolutionary psychologists have tested this hypothesis by asking women of means (medical students) what kind of man they might want. These women stated that financial earning power equal to, or better than, their own was important to their mate choices (Townsend 1987; Townsend 1989). One obvious criticism of this work is that adult women have already been socialized to conform to societal standards and find a man even more successful than themselves. What would be of interest is a follow-up study of these women to see whom they actually marry.
18. Duck and Miell 1983.
19. Singer 1985.
20. Rosenblatt and Anderson 1981.
21. Rosenblatt and Anderson 1981; Frayser 1985; Buss 1994.
22. Frayser 1985.
23. Broude and Greene 1976.
24. Buss and Schmitt 1993; Buss 1994.
25. Small 1992a; Small 1993.
26. Broude and Greene 1976; Whyte 1978; Rosenblatt and Anderson 1981; Small 1993.
27. Diamond 1992; Buss 1994.
28. Baker and Bellis 1993a.
29. Frayser 1985.
30. Pusey and Packer 1987.
31. Goodall 1986.
32. Pusey and Packer 1987.
33. Crockett 1984.
34. Frayser 1985.
35. Malinowski 1965.
36. Lévi-Strauss 1969.
37. Although this study points to a proximate mechanism of familiarity breeding contempt for the ultimate ends of incest avoidance, it's also possible that the organizational rules of the community and the need to diffuse any possibility of internal community competition, as Malinowski suggested, are the source of members not marrying (Talmon 1964). However, other groups, such as the Muria in India, let youngsters live side by side in adolescence and marriages are rarely made between those from the same dormitory (Rosenblatt and Anderson 1981).
38. Fox 1980; Frayser 1985.
39. Frayser 1985.
40. Rosenblatt and Anderson 1981.
41. Whyte 1978.
42. Frayser 1985.
43. Fisher 1992; Small 1992b.
44. Small 1992b; Small 1993.
45. Frayser 1985.
46. Frayser 1985; Small 1992b.
47. Frayser 1985.

48. Rosenblatt and Anderson 1981; Small 1993.
49. Ellis 1992; Jackson 1992.
50. Trivers 1972.
51. Symons 1979; Buss 1994.
52. Ford and Beach 1951.
53. Ford and Beach 1951.
54. Ford and Beach 1951.
55. Buss 1994.
56. Dion 1981.
57. Ford and Beach 1951.
58. Langlois and Roggman 1990.
59. Johnson and Franklin 1993.
60. Cunningham et al. 1990.
61. Perrett et al. 1994.
62. Bromberg.
63. Fallon and Rozin 1985; Rozin and Fallon 1988.
64. Singh 1993.
65. Buss 1994.
66. Hickling et al. 1979; Wilson 1981; Nevid 1984; Buss 1985; Carroll et al. 1985; Buss 1986; Howard et al. 1987; Buss 1989; Townsend and Levy 1990; Buss 1994. These differences also hold up for homosexual individuals (Jankowiak et al. 1992).
67. Hickling et al. 1979.
68. Townsend 1987; Townsend 1989.

69. Wiederman and Allgeier 1992.
70. Harrison and Saeed 1977; Deaux and Hanna 1984; Thiessen et al. 1993.
71. Gladue and Delaney 1990; Kendrick et al. 1990; Kenrick et al. 1993.
72. Buss 1994.
73. Symons 1993.
74. Buss 1994.
75. Buss 1989.
76. Irons 1983; Lancaster and Lancaster 1983; Lancaster and Lancaster 1987; Fisher 1992; Small 1993.
77. Duck 1973; Duck and Miell 1983.
78. Greiling 1993.
79. Small 1992a; Small 1993.
80. Buss 1985; Buss and Schmitt 1993.
81. Berscheid et al. 1971; Berscheid and Wlaster 1974.
82. Murstein and Christy 1976; Price and Vandengerg 1979.
83. Kendrick et al. 1993.
84. Eckland 1968; Michael et al. 1994.
85. Price and Vandengerg 1979; Buss 1989.
86. Whyte 1990; Michael et al. 1994.
87. Dickemann 1979; Frayser 1985.
88. Rushton 1988.
89. Duck and Miell 1983.
90. Murstein 1976; Nevid 1984.
91. Betzig 1989; Fisher 1992.

CHAPTER SIX

1. Money 1991.
2. Money and Ehrhardt 1972.
3. Bem 1993.
4. Bem 1993.
5. Ford and Beach 1951.
6. Fay et al. 1989; Weeling et al. 1990; France 1992; Johnson et al. 1992; Diamond 1993; Muir 1993; Turner 1993; Michael et al. 1994.
7. Fay et al. 1989; Diamond 1993; Janus and Janus 1993; Michael et al. 1994.
8. Diamond 1993.
9. Kinsey et al. 1948.
10. An eloquent essay on the continuum of human sexuality appeared in *The Nation* on October 19, 1992. D. L. Rist points out that exclusively homosexual or heterosexual, the two ends of the continuum, is rather perverse. Each of us, even

the most homosexual or heterosexual, has probably had an occasional fantasy about the sex we've never experienced.
11. France 1992; Johnson et al. 1992; Diamond 1993.
12. Kluckholn 1948.
13. Whitehead 1981.
14. Ford and Beach 1951.
15. Ford and Beach 1951.
16. For example, the analysis of homosexuality in Mombasa by Shepard explains the possibility of homosexuality as a way to improve one's status and wealth (Shepard 1987).
17. Carrier 1980. This approach to homosexuality is one of the problems with AIDS education in some Latino cultures. If a man has sex with another man, but he is

the inserter and therefore not "homosexual," how can he be at risk for AIDS?
18. Whitehead 1981; Williams 1986.
19. Williams 1986.
20. Nanda 1986; Nanda 1991.
21. Ford and Beach 1951.
22. Callender and Kochems 1986.
23. Carrier 1980.
24. Adam 1986.
25. Herdt 1981; Stoller and Herdt 1985.
26. Kinsey et al. 1948; Ford and Beach 1951.
27. Shepard 1987.
28. Bem 1993.
29. Bancroft 1980.
30. Isay 1989; Bem 1993.
31. Isay 1989; Bem 1993.
32. Bieber et al. 1962.
33. Bayer 1987.
34. Bayer 1987; Isay 1989.
35. Hooker 1965.
36. Isay 1989.
37. Isay 1989.
38. Hamer et al. 1993; Pool 1993; LeVay and Hamer 1994.
39. Pillard and Weinrich 1986.
40. LeVay and Hamer 1994.
41. Kallmann 1952; Heston and Shields 1968; Byne and Parsons 1993.
42. Elke et al. 1986.
43. Bailey and Pillard 1991.
44. Bailey et al. 1993.
45. Whitam et al. 1993.
46. Turner 1994.
47. Byne and Parsons 1993; Byne 1994.
48. Young et al. 1964.
49. Dörner 1976; Dörner 1988.
50. Young et al. 1964; Gorski et al. 1978.
51. Byne and Parsons 1993; Byne 1994.
52. Byne and Parsons 1993.
53. Meyer-Bahlberg 1984; Adkins-Regan 1988.
54. Bell et al. 1981; Dörner 1976; Dörner 1988.
55. Money and Ehrhardt 1972.
56. Money and Ehrhardt 1972.
57. Imperato-McGinley et al. 1979.
58. Ehrhardt and Meyer-Bahlberg 1981.
59. Meyer-Bahlberg 1984.
60. Money 1988.
61. Persky 1987.
62. Ellis et al. 1988; Bailey et al. 1991.
63. Money and Ehrhardt 1972.
64. Dörner et al. 1975.
65. Gladue et al. 1984.
66. Byne 1994.
67. Meyer-Bahlberg 1984; Gooren 1990.
68. Gooren 1986.
69. Meyer-Bahlberg 1984.
70. Kimura 1992.
71. Gorski et al. 1978.
72. Swaab and Fliers 1985; Allen et al. 1989; LeVay 1991; Byne 1994; Byne in press.
73. Baringa 1991; Ezzel 1991; LeVay 1991.
74. Allen and Gorski 1992; Byne and Parsons 1993; Byne 1994.
75. Swaab and Hofman 1990.
76. Bancroft 1980.
77. Adkins-Regan 1988; Byne and Parsons 1993; Byne 1994; Byne in press.
78. Byne in press.
79. Nadler 1990.
80. Smuts and Wantanabe 1990.
81. Kano 1992.
82. Carpenter 1942; Chevalier-Skolnikoff 1976; Akers and Conaway 1979; Wolfe 1979; Wolfe 1984; Wolfe 1986; Sommer 1988; Srivastava et al. 1991; Kano 1992.
83. Small 1993.
84. Fedigan and Gouzoules 1978; Wolfe 1979; Gouzoules and Goy 1983; Wolfe 1984; Wolfe 1986.
85. Sommer 1988; Srivastava et al. 1991.
86. Thompson-Handler et al. 1984; Kano 1992.
87. Wilson 1978.

CHAPTER SEVEN

1. Nevid 1993.
2. Control 1991.
3. Altman 1986; Grmek 1990.
4. Grmek 1990.
5. Nevid 1993.
6. Miller et al. 1990.
7. Voeller 1991.
8. Miller et al. 1990.
9. Altman 1986.
10. Boswell 1990.

11. Altman 1986.
12. Michael et al. 1994.
13. Grmek 1990.
14. Stall et al. 1989; Smith 1991.
15. Janus and Janus 1993.
16. France 1992; Turner 1993.
17. France 1992.
18. Control 1990.
19. Pfeffer 1993.
20. Insler and Luenfled 1993.

21. Collins et al. 1983.
22. Sandowski 1993.
23. Strathern 1992.
24. Steinem 1994.
25. Lefrançois 1993.
26. Hahn and Stout 1994.
27. Elmer-Dewitt 1994.
28. Tierney 1994.
29. Hein 1993; Wexelblat 1993.

Index